EQUIPMENT PLANNING FOR TPM

Maintenance Prevention Design

EQUIPMENT PLANNING FOR TPM

Maintenance Prevention Design

Fumio Gotoh

Originally published by the Japan Institute
for Plant Maintenance

Productivity Press, Inc.
Cambridge, Massachusetts Norwalk, Connecticut

Originally published in Japanese as *Setsubi kaihatsu to sekkei* by the Japan Institute of Plant Maintenance, Tokyo, Japan. Copyright © 1988.

Productivity Press, Inc.
P.O. Box 3007
Cambridge, MA 02140
(617) 497-5146

Library of Congress Catalog Card Number:
ISBN: 0-915299-77-1

Cover design by David B. Lennon
Printed and bound by Maple-Vail Book Manufacturing Group
Printed in the United States of America

Library of Congress Cataloging-in-Publication Data

Gotō, Fumio.
[TPM no tame no setsubi kaihatsu to sekkei. English]
Equipment planning for TPM : maintenance prevention design / by Fumio Gotoh.
p. cm. — (Total productive maintenance series)
Translation of : TPM no tame setsubi kaihatsu to sekkei.
Includes index.
ISBN 0-915299-77-1
1. Plant maintenance—Management. 2. Industrial equipment—Maintenance and repair—Management. I. Title. II. Series.
TS192.G6813 1991
658.2'02—dc20 91-6737
CIP

91 92 93 10 9 8 7 6 5 4 3 2 1

Contents

NOTE: Due to the format requirements of the illustrations, many of which are two-page spreads, you will occasionally find a blank page. Please note that these are intentional, and that there is no material missing from the book.

Publisher's Message

Just three years ago, Seiichi Nakajima's overview of total productive maintenance was published in English. At that time very few American manufacturing managers and engineers had even heard of the companywide program Japanese firms were using to dramatically reduce equipment-related costs. Today, the basic principles of TPM are more widely understood. A growing number of companies are developing implementation strategies to establish TPM successfully in the American manufacturing environment. Many companies are developing team approaches to improve preventive maintenance, raise the performance of equipment, lengthen mean time between failures, and break down long-standing barriers between production and maintenance personnel. They are also using condition-based maintenance increasingly for more precise and cost-effective predictions of the period between failures.

These types of improvement activity can dramatically increase equipment availability and reliability and lengthen equipment's valuable life. At the same time, however, the root causes of equipment life cycle costs in planning and design must be addressed. Up to 75 percent of life cycle costs are attributable to operation and maintenance, but most of these are the result of decisions made at the design stage. Excellent preventive maintenance combined with judicious modifications for reliability and maintainability can reduce running costs somewhat. But today's shop floor improvement must be carried out with one eye focused on the opportunity for further cost reduction through better equipment designs.

In *Equipment Planning for TPM: Maintenance Prevention Design*, author Fumio Gotoh shows us how the Maintenance Prevention (MP) design information gathered in TPM is used in design activities and demonstrates why that data gathering is so important — even for companies that do not currently design their own equipment in house. Drawing on his long experience with companies like Toyota and its affiliates, Gotoh offers us a unique insight into how top Japanese companies have dealt successfully with equipment planning issues that originally surfaced as quality, maintainability, or flexibility problems on the shop floor. He describes in elaborate detail how information about existing equipment can be analyzed and used to develop more effective designs that meet quality, reliability, cost, and flexibility goals. He also suggests how we can manage the process from design to installation in order to shorten development time significantly and eliminate or minimize the costs, delays, and other headaches associated with the period from installation to full-scale production.

Why has the approach detailed here been so successful in Japan? First, a consistent clarity about the purpose of the design activity is shared among team members from the outset. This clarity is maintained through adherence to a careful and thoroughly planned design process. Moreover, as in other TPM activities, timely, effective communication and cooperation are repeatedly emphasized. Team members' efforts are further supported by working charts and diagrams that help organize and analyze information, as well as plan and manage follow-up activities. I encourage you to study the sample management charts included in each chapter. They represent excellent applications in equipment design of the "seven new tools" of quality management.

Second, the activity is carried out by concurrent engineering teams from design, manufacturing, and maintenance departments — in a process designed to ensure that issues relevant to those often competing perspectives are thoroughly addressed when and as they come up. While not the subject of this book, concurrent or simultaneous engineering is the strategy driving the processes Gotoh prescribes. These processes reveal some of the ways in which Japanese manufacturers have been reducing time-to-market so dramatically.

Finally, whereas TPM activities on the shop floor are dedicated to the elimination of equipment-related waste, the activities described in this book are dedicated to waste *prevention*. Whether by reducing initial and running costs, preventing quality problems through "QA-friendly" features, or by designing in ease of operation, adjustment, and servicing, MP design represents a more sophisticated stage of equipment-consciousness — one which we should begin to internalize even as we pursue fundamental shop floor strategies.

This book is the fourth in our Total Productive Maintenance Series; it represents a turning point in emphasis as we bring into sharper focus different methodologies associated with the application of TPM. Our thanks go to Bruce Talbot, who patiently translated the book, and to Michael Kelsey and Drew Dillon, who helped clarify some of the more difficult technical passages. Special thanks are also due to Christine Reynolds and our production staff of Gayle Joyce, Susan Cobb,

Caroline Kutil, Gary Ragaglia, and Karla Tolbert who painstakingly developed new designs for the original management charts and diagrams, and to Production Manager David Lennon for project management and the cover design.

Norman Bodek
President

Connie Dyer
Senior Editor

Foreword

Japan's PM Prize for excellence in preventive and productive maintenance was first awarded in 1964. Since then the field of industrial equipment management has been gradually transformed and enriched. Today TPM is gaining attention among industrial circles worldwide as it continues to grow and develop. No other management technique has contributed more to the reputation for reliability that Japanese manufacturing industries currently enjoy.

Each year — with a few notable exceptions — almost every prize-winning company's TPM activities focus on the five fundamental TPM improvement activities — autonomous maintenance, equipment improvements, maintenance planning, education and training, and early equipment management. Nevertheless, something seemed to me to be missing. PM prize judges have spent most of their time studying improvements and large-scale QC circle activities, and the connection between maintenance and equipment and process planning has not been explored in any depth.

Obviously, improvement is an integral part of TPM, but what could be more important than the long-range planning that guides the factory's activities? If this function is inadequate, there will be weaknesses in plans that support maintenance activities (including maintainability improvements and maintenance prevention) and in plans for the production equipment itself (e.g., performance, cost, safety, and other related equipment design factors). This will be true whether the equipment is developed and manufactured in-house or purchased from outside vendors. In my opinion, many manufacturers suffer from this imbalance — focusing too much on improvement and not enough on equipment planning.

There are various reasons for this imbalance, one being the lack of books specifically addressing the issue. Thanks to Mr. Gotoh, this book offers a long-needed focus on this vital topic and provides a treasure trove of theory and case studies to clarify the issues.

Once every few years Japan's TPM movement reaches a turning point that adds new dimensions. When this book was written we were at such a turning point, and the new dimension was planning. The timing of this book could not have been better.

In Japan, "early equipment management" generally refers to the sequence of management activities aimed at getting new equipment installed and operating normally and stably. In this book, however, Mr. Gotoh adopts a broader definition of early equipment management. Using case studies and lucid explanations of sophisticated concepts, he covers various aspects of design for TPM, including design for quality assurance, design to life cycle cost, and design for flexibility and cost-efficient automation — all to increase quality, reduce development lead times, lower costs, and guarantee safety.

Fumio Gotoh has brought to this book the expertise and authority gained during his eighteen-plus years as a PM consultant for over fifty firms. His experience as an IE and QC consultant has also helped to broaden his perspective. I first met Mr. Gotoh during his college days when he worked in my research center at Keio University. He has since built a long and successful career as a consultant specializing in TPM. The length and depth of his experience in this field are evident in this book's wealth of practical advice and explanations.

One final acknowledgment: Mr. Gotoh enlisted the cooperation of some planning powerhouses, such as Toyota Auto Body, in conducting the innovative experimental projects he reports in detail. This cooperative effort not only enriches the book, it also provides the impetus for dramatic TPM achievements by companies everywhere.

Shizuo Senju
Professor Emeritus, Keio University

Preface

Since the Japan Institute for Plant Maintenance (JIPM) published *TPM Development Program*, known as the TPM "bible," many companies have applied its lessons with great success. As one of the book's five authors, I have enjoyed witnessing their achievements.

TPM Development Program brought together the five TPM development activities — autonomous maintenance, equipment improvements, maintenance planning, education and training, and early equipment management — emphasizing their integration and approach. The book also provided practical advice on how they should be implemented, except in the area of early equipment management, where a largely theoretical discourse left many readers wishing for more practical advice. These wishes are being answered now, thanks to the many helpful people who assisted in producing this book.

One other background note concerns the increasingly close relationship between corporate competitiveness and equipment development capacity. Today, companies must continue to grow in an environment where competition and consumer needs change at an accelerating pace. To develop and maintain the organizational and technological strength to keep pace with these changes requires advanced product and equipment development programs.

The two greatest challenges for future product and equipment development will be shortening lead time from development to production and building higher quality into both products and production equipment. Consequently, all technical fields related to equipment design and development will become increasingly important in coming years.

I organized this book around early equipment management activities aimed at reducing lead times and around engineering activities that promote higher quality. Management activities that can be improved to reduce lead time include:

- the approach to product development and design,
- setting of design goals,
- preliminary evaluation,
- cost-effective automation design, and
- safety assurance design.

While this book consciously distinguishes between management and engineering activities, readers should understand that these spheres are interdependent and complementary.

We must find ways to promote — systematically — equipment development and design processes that are thorough as well as efficient. Companies that tackle technical challenges before establishing reliable processes are likely to encounter problems arising from lack of technical data, time constraints, lax deadlines, and lack of criteria for setting priorities. Companies that do not resolve these problems will find it difficult to manage successfully the five activities listed earlier.

The purpose of this book is to help companies advance technologically while also managing successfully. Companies that fail to tackle both challenges together will have much more difficulty solving technical issues; they will also miss an opportunity to build a powerful corporate base resulting from dramatic improvement in both engineering performance and organizational strength. This book is intended to provide the means for building that base.

This book focuses on just one of the five major TPM development activities — early equipment management — and is thus intended for engineers involved in production equipment design and development, specifically production engineers. It may also be useful to maintenance technicians who routinely give feedback to production engineers concerning equipment improvements, MP data, and other matters, and to engineers involved in product development (capital-goods products) who are looking for ways to improve their approach to production development and design.

I also recommend that managers read at least the first chapter for an overview of early equipment management. In this book, I have emphasized two pressing issues: training for engineers and ways to strengthen the ties in manufacturing among product development, equipment design, and maintenance divisions. The issues and challenges presented here should help engineers change the way they think about their work.

Readers who want to apply the problem-solving approaches described in this book should note the following points. The effectiveness of any method will vary depending on the type of company where it is applied. Consequently, readers should understand clearly their own company's problems and goals before attempting to apply any of the approaches described in this book. Since the five managerial issues discussed in Chapter 2 are resolved by a more general approach, I suggest that you tackle these problems first before addressing the engineering-related issues.

Each problem-solving method is explained conceptually and practically. To facilitate comprehension, I have included practical case examples. Do not assume, however, that the approach described in the example will work exactly the same way in your own company. Simply copying the approach without understanding may result in frustration that can undermine your overall TPM efforts. First grasp the basic principles behind the method, then you can adapt it more easily to suit your company's individual needs.

For their invaluable help in getting this book to press I thank first my many friends at Toyota Auto Body, especially director Akira Iijima and production engineering division chief Mitsuo Sakaguchi, who with others in the production engineering division helped collect data for the case studies. I am also grateful for the case studies provided by Ishikawa Steel, Chuo Hatsujo, Tokico, Topy Industries, and Toyoda Automatic Loom Works.

I am also deeply indebted to several people for their editorial assistance, to Dr. Zentaro Nakamura of Keio University's science and engineering department, who helped with the entire manuscript, and to Mr. Etsusaburo Murakami of the Bridgestone Technical Center for his suggestions and guidance concerning intrinsic safety design.

Finally, I offer my heartfelt gratitude to Mr. Soneda, publishing division chief at JIPM, for all his kind assistance.

Fumio Gotoh

1
Overview of Early Equipment Management

During Japan's high-growth era, most Japanese companies experienced relatively smooth growth and expansion. Many received top marks worldwide for quality and profitability. The business environment around the world is growing harsher, however, and today's successful companies have no guarantee they will continue to thrive in the future.

NEW ISSUES FOR TODAY'S COMPANIES

Now that many of the products manufactured by leading Japanese companies have reached maturity, these companies are striving to respond to the needs of mature, stable-growth markets. Meanwhile, new environmental factors, such as the dampening effect of the high yen on Japanese exports and the challenge posed by newly industrialized economies (NIEs), make it increasingly difficult for companies to remain profitable. Companies in Japan and elsewhere must address a new set of problems posed by a harsher business environment.

First, the product development picture is changing: product development is being polarized into high-grade, high-priced products and lower-grade, low-priced products. The need for both types has significantly increased the number of product models. At the same time, companies are diversifying into new or different product fields, or both. Second, many manufacturing companies are being forced to set up factories overseas to keep labor costs on a par with those of the NIE competitors. Third, these fast-paced changes require equally swift responses. Companies lagging even a little behind the pace of change in the economic environment can easily lose the edge to fast-moving competitors. As competition becomes increasingly time-based, companies will need to speed up their analysis of current conditions to make more timely responses.

TECHNICAL ISSUES

In view of these three factors, what technical issues must companies address to keep pace with the changing environment?

Product development. In addition to developing appealing products, manufacturers must design products that are easier to produce — by cutting production costs, facilitating automation, shortening development and production lead times, stabilizing quality, and facilitating overseas production. When companies pursue a strictly market-driven product design policy, products are harder to produce, their quality is less even, cost reduction is hard to accomplish, and overseas expansion becomes problematic.

Production engineering. Production engineering for the future should generate new methods for building equipment that are flexible enough to produce diverse product models economically. It should also result in low-cost automation, optimal equipment life cycle costs, greater equipment precision, and more innovative, detailed equipment designs.

Manufacturing technology. Companies need the technological base to develop more automated, sophisticated, and advanced manufacturing processes, and they must ensure a steady flow of feedback on current problems relating to design and technology. This cannot be done by equipment operators, no matter how well trained. Companies must also cultivate a staff of engineers who are highly skilled in crucial high-tech areas.

This suggests that future total productive maintenance (TPM) development activities must be broad enough in scope to clarify and resolve the substantial issues related to product design, production engineering, and manufacturing. It may be difficult to envision just how making substantial improvements in the narrow field of factory operations will enable companies to respond effectively to the environmental issues described here. Figure 1-1 clarifies the relationship between those issues and TPM.

EARLY EQUIPMENT MANAGEMENT IN TPM

Many Japanese companies have introduced TPM development programs. Between 1971 and 1989, 171 companies won the PM Prize* by applying these programs effectively. TPM helped them perform well in today's slower-growth business environment. In the coming years, TPM programs will become even more important for corporate survival as the business environment grows harsher.

Outlook for TPM

Companies that want to survive in today's highly competitive business environment must address the need for high

* The Japan Institute for Plant Maintenance (JIPM) sponsors the annual PM Prize for Excellence in TPM.

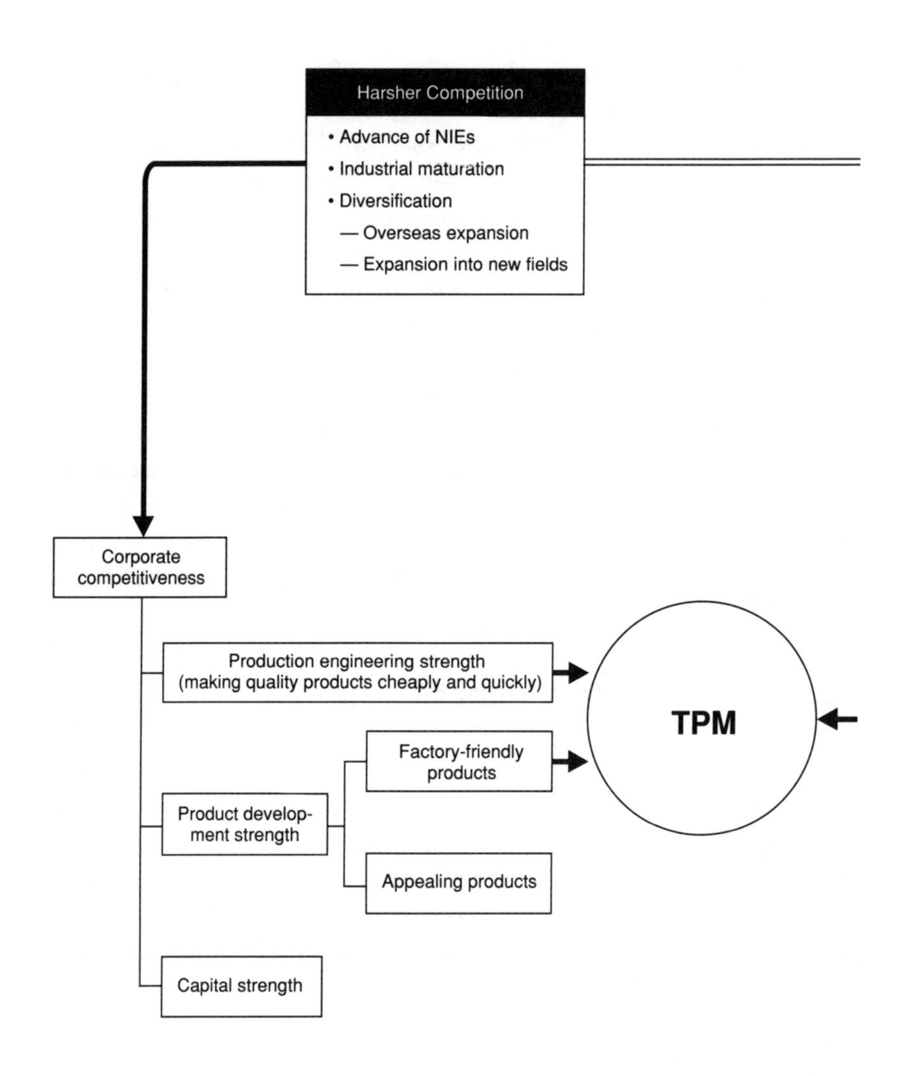

Figure 1-1. TPM in a Changing Business Environment

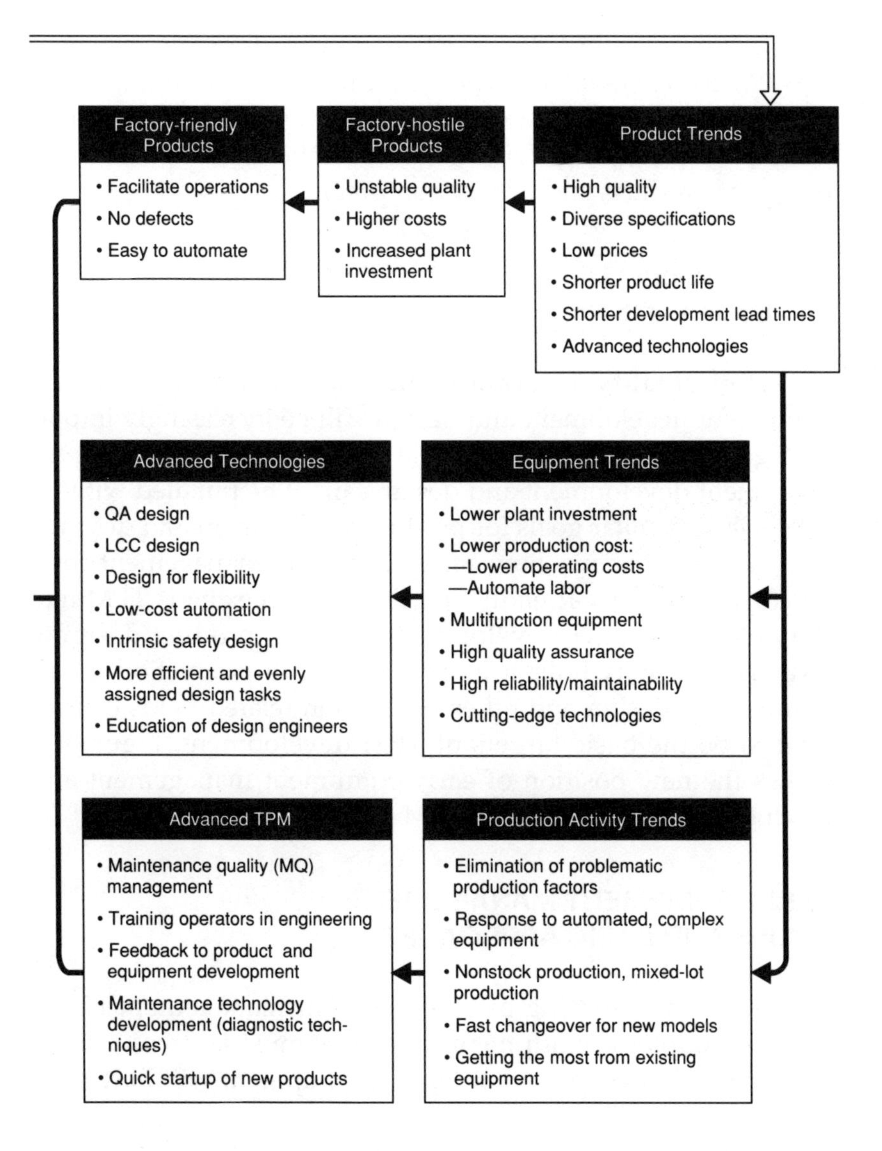
Factory-friendly Products
• Facilitate operations
• No defects
• Easy to automate
Factory-hostile Products
• Unstable quality
• Higher costs
• Increased plant investment
Product Trends
• High quality
• Diverse specifications
• Low prices
• Shorter product life
• Shorter development lead times
• Advanced technologies
Advanced Technologies
• QA design
• LCC design
• Design for flexibility
• Low-cost automation
• Intrinsic safety design
• More efficient and evenly assigned design tasks
• Education of design engineers
Equipment Trends
• Lower plant investment
• Lower production cost:
—Lower operating costs
—Automate labor
• Multifunction equipment
• High quality assurance
• High reliability/maintainability
• Cutting-edge technologies
Advanced TPM
• Maintenance quality (MQ) management
• Training operators in engineering
• Feedback to product and equipment development
• Maintenance technology development (diagnostic techniques)
• Quick startup of new products
Production Activity Trends
• Elimination of problematic production factors
• Response to automated, complex equipment
• Nonstock production, mixed-lot production
• Fast changeover for new models
• Getting the most from existing equipment

quality, lower costs, and more effective, swifter research and development (R&D). How should a company's program change to survive in this tougher environment? One way is to take TPM development beyond the division-specific programs proposed in *TPM Development Program* (e.g., for the maintenance and operations divisions), to emphasize TPM's potential contributions in engineering departments, in the development and design of products and factory equipment.

Linking Product Development and Design with Production

High quality assurance combined with lower costs in equipment development and design will be increasingly important goals in years to come. Naturally, these twin goals for equipment development and design cannot be pursued without establishing similar goals for product development and design.

Figure 1-2 shows how early equipment management combined with "factory-friendly" product design extends TPM into the field of product development. Making product design a sixth TPM activity links it more closely with production development and design and other production-related fields considered to be the basic targets of TPM development. Figure 1-2 shows the new position of early equipment management and product design in the overall TPM picture.

EARLY EQUIPMENT MANAGEMENT: ISSUES AND BASIC APPROACHES

Before considering early equipment management issues and methodology, equipment managers must address a more fundamental question: what kind of equipment will best meet future needs?

Equipment is becoming an essential ingredient for competitiveness. Advances in automation and mechatronics are mak-

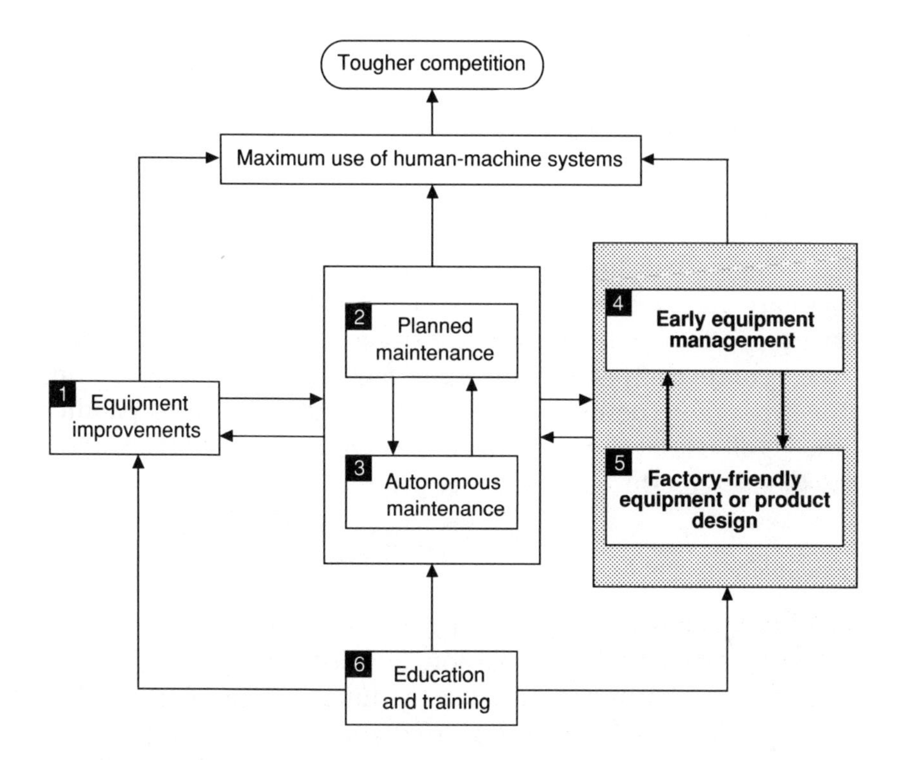

Figure 1-2. The New Position of Early Equipment Management in TPM

ing factory equipment a more decisive factor in determining product quality, volume, and cost. Equipment must be flexible to respond to abrupt changes in market demand as well as specialized to help the company remain competitive.

Five Conditions for Early Equipment Management

The following basic conditions for factory equipment of the future are related to the various requirements posed by emerging environmental factors:

- Development
- Reliability

- Economics
- Availability
- Maintainability*

Development

Demand for higher quality and growing price competitiveness are creating new needs related to fabrication and equipment development. Today, R&D strength is a key determinant of corporate competitiveness. True competitiveness comes from the synergetic combination of product development capability and rigorous development of equipment and related fabrication methods to produce high-quality products at low cost.

Reliability

The equipment of the future will be novel in many respects. As a rule, however, more novelty means less reliability. New equipment that is only slightly different from older models benefits from the inclusion of tried-and-true technologies that make for high reliability. Substantially new equipment designs, however, include experimental devices or technologies almost by definition and are likely to require some debugging after installation and startup. The experimental nature of new equipment is precisely what makes post-installation problems much more likely. The challenge, therefore, is to create radically new equipment while accurately estimating beforehand what problems may occur and using preliminary evaluation methods (such as those described in Chapter 3) to prevent their occurrence.

* The initial letters of these five words spell DREAM, an appropriate acronym for the industrial "dream machines" that early equipment management seeks to create.

Economy

Manufacturing companies use factory equipment to build products that must be sold for a profit. This principle of manufacturing economics is the same the world over and cannot be avoided. For example, certain new equipment fabrication methods might turn out extremely reliable equipment. If fabrication costs are too high, however, the equipment will not be profitable to produce and will likely end up as scrap metal.

Progress and change are obvious. Companies can no longer enjoy the luxury of suffering short-term losses to build up technologies for the future. Industrial equipment developers cannot afford to develop the finest, highest-performance equipment. Instead, they must balance performance with economy by building the least expensive equipment that fulfills all the required functions.

Equipment must be inexpensive not only in terms of initial costs (procurement costs, fabrication costs, etc.) but also in terms of its running costs (operation and repair expenses, etc.). This brings us to the key concept of life cycle cost (LCC) — which combines initial costs and running costs. The industrial equipment developer's object is to develop and design equipment with the lowest possible life cycle cost (see Chapter 4).

Availability

Availability can be defined as the ratio of fully functional equipment to all reparable (functioning and nonfunctioning) equipment within a given period of time. For most equipment, the three major obstacles to availability are:

- Breakdowns
- Downtime for maintenance
- Changeovers

The first obstacle is a reliability issue and the second is a maintainability issue.

If we broaden the definition of availability, we can interpret it as a measure of effectiveness in capital investment. Thus,

- Availability drops when equipment is left idle as a result of reductions in production, cycle time imbalances, or other factors.
- Availability is also reduced when equipment is left idle because of product changes that result in product/equipment incompatibility.

These types of idle time along with changeovers are related mainly to changes brought about by market trends. Since the outlook is for even faster-changing markets, these latter three factors will probably become more significant.

Design for flexibility (the subject of Chapter 5) seeks to avoid obstacles to availability, including those resulting from product diversification and variable production yields. Equipment designs must flexibly accommodate changes in the business environment as well as in the company's internal operations. To do otherwise in these fast-changing times invites equipment obsolescence even before the company can recoup its equipment investment costs. Since equipment investment costs are often very high, such early obsolescence can deal a fatal blow to some companies.

Maintainability

Maintainable equipment does not break down easily or often and is easy to repair when maintenance problems do occur. The second, third, and fourth requirements (reliability, economy, and availability) all require the support of proper maintenance. Spending too much money on maintenance, however, works against the economy requirement, and spending too much time on maintenance interferes with the availability requirement.

Generally, maintainability requires a balance between two kinds of costs. On one hand, we must strengthen preventive maintenance to minimize the loss-related costs of equipment downtime. On the other hand, we cannot raise investment costs too much in support of preventive maintenance.

Improving maintainability keeps costs from increasing on either side of this balance point. One way to improve maintainability is to modify existing equipment to make it easier to maintain. But this after-the-fact approach is not nearly as effective as building in maintainability at the equipment-design stage. A maintenance prevention (MP) design approach (discussed on pages 177-195) seeks to minimize post-installation problems in equipment by making it highly maintainable from the start. MP design is one of the key elements in life cycle cost (LCC) design, discussed in greater detail in Chapter 4.

Overview of Early Equipment Management

Figure 1-3 is a schematic overview: the left side of the figure outlines the procedure for carrying out early equipment management; on the right are technical issues in equipment development and design that are addressed through the procedure in a timely and effective manner. The activities addressed through early equipment management include (1) initial development and design of products and equipment, (2) the sequence of activities from process planning to initial operation and production-related stages (the main target of early equipment management), and (3) developers' use of feedback from the factory.

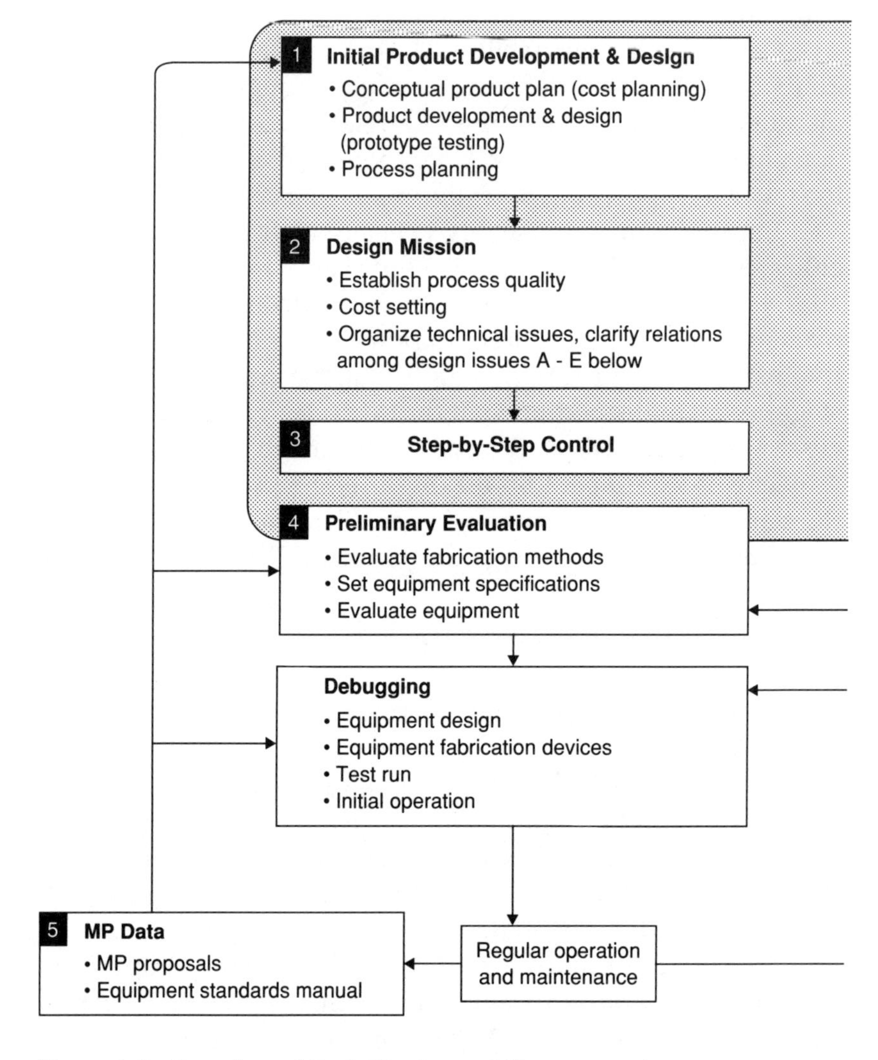

Figure 1-3. Overview of Early Equipment Management

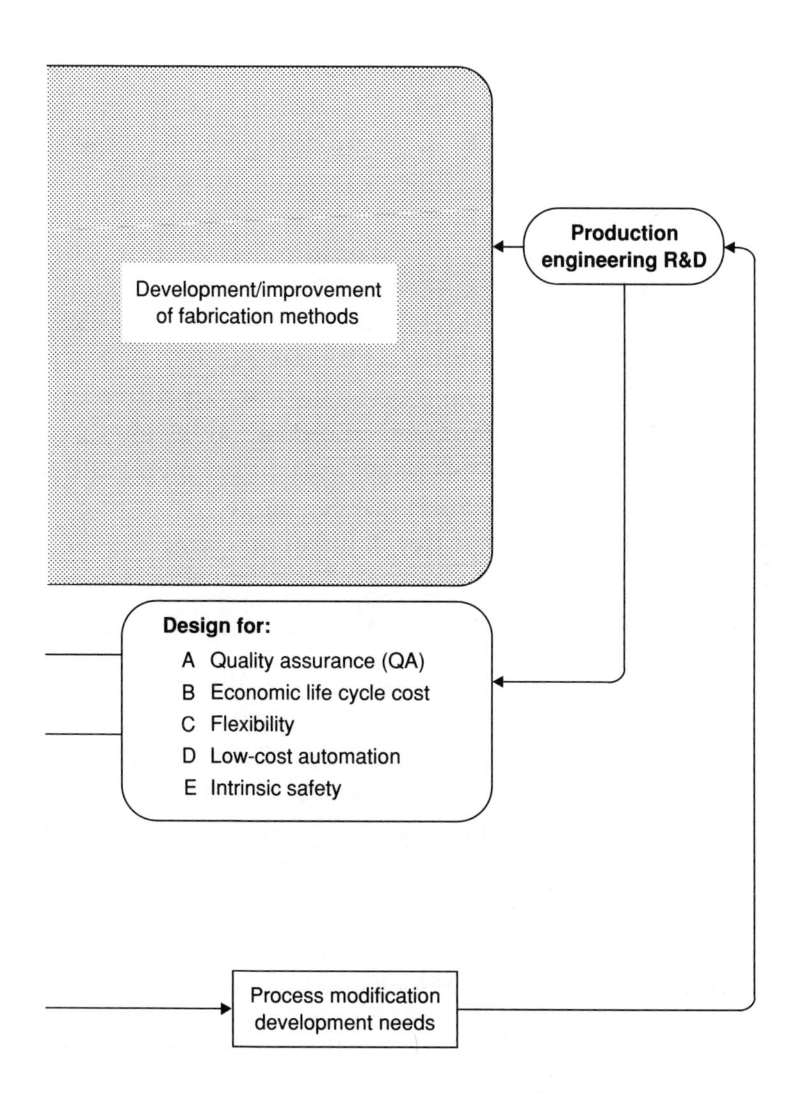
Production engineering R&D
Development/improvement of fabrication methods
Design for:
A Quality assurance (QA)
B Economic life cycle cost
C Flexibility
D Low-cost automation
E Intrinsic safety
Process modification development needs

Early Equipment Management Objectives

Early equipment management has three main objectives. The first is to fulfill 100 percent of the quality characteristic values requested by the product designers. To do this, we must establish process quality in the process planning stage and then build equipment that is able to provide 100 percent of the process quality.

The second objective is to ensure that the planned production capacity is based on the product's cost plan while establishing target values for the initial cost and running cost of each process concerned. Then we will know exactly what is required of the equipment and can procure the least expensive equipment that meets 100 percent of the requirements at the lowest cost.

The third objective is to get the equipment up and running reliably so that the products can be shipped according to schedule and the equipment is free of initial problems.

Technical Issues in Early Equipment Management

The main technical issues are listed in Figure 1-3. Addressing these issues assures equipment that promotes optimal human-machine relationships, higher quality assurance, significantly lower life cycle costs, and less costly, intrinsically safer automated equipment. These issues are addressed in the early equipment management process (Steps 1 through 5 in Figure 1-3).

Common Approaches to Effective Management

Typically, there are four approaches to managing the issues just described. The first is to establish precise (quantified) design goals for both quality and cost. The developer or designer's most important task is to find ways to meet these design goals.

A second approach is to consider equipment development and design at the product development and design stage. Once a product has been designed, the range of possible equipment design elements is limited and major improvements in equipment design are much more difficult to achieve.

A third approach is to address equipment problems that have been identified in the factory. We must acquire a clear understanding of the problems and other conditions on the shop floor and incorporate the solutions into new equipment development and design.

The fourth approach is to conduct thorough preliminary evaluation and debugging of the equipment during the development, design, and post-installation stages. The goal is to anticipate problems and prevent their occurrence at the earliest step in the development and design ladder. The earlier this is done, the less damage will be incurred. Costs will be lower, lead times shorter, and startup operations less likely to be delayed.

Early Equipment Management Issues

The early equipment management issues listed in Figure 1-3 are described below in detail.

Design for Quality Assurance

When new equipment is installed, defects typically show up one after another until the whole factory floor lapses into chaos. Modification follows modification until, finally, the equipment appears to be turning out products properly. Before long, however, precision deteriorates until a new set of defects occurs and must be eliminated.

QA design addresses this issue by seeking to ensure 100 percent defect-free equipment beginning at the design stage. The QA design approach has two goals, the first of which is equipment free of initial defects. This means designing equipment with functions that guarantee defect-free operation from initial startup.

The second goal is QA-friendly equipment. Designing equipment guaranteed to operate defect-free is futile if that defect-free operation cannot be counted on to last. In other words, we must design equipment to hold tolerances over time and to be resistant to defects resulting from random disturbances. What this amounts to is QA design for quality that lasts for the entire service life of the equipment. The specific steps for establishing a QA design are described in Chapter 3.

Design to Life Cycle Cost

Until recently, most equipment developers and designers emphasized equipment's initial costs (design and procurement costs) over its running costs. As running costs have escalated, more developers and designers are thinking in terms of the total cost or life cycle cost (LCC). They have learned that the surest path to profitability lies in minimizing LCC. LCC design refers to a design approach focused on minimizing LCC.

There are four main LCC design strategies, each oriented toward different equipment characteristics or environmental conditions.

Minimum initial cost (IC) design. This approach concentrates on minimizing the initial cost of the equipment. For example, if the initial cost element of the LCC for a press die is extremely large, the running cost is less important, and minimum IC design becomes the most effective design approach.

Minimum running cost (RC) design. In other cases, it is best to emphasize reducing the running cost over reducing the initial cost. In complicated automated equipment, higher quality generally means greater risks of defects, breakdowns, erratic starts and stops, and other problems. It also leads to more laborious maintenance and more expensive spare parts. As a result, running costs such as downtime loss and maintenance costs soon exceed their initial projections. In such cases, more emphasis on reliability and maintainability will minimize running cost.

IC-RC reduction design. This approach gives equal emphasis to both kinds of costs. Obviously, this amounts to a holistic LCC reduction approach. For example, painting equipment tends to have high initial costs *and* high running costs (paints, energy, defect-related loss, etc.). The ICR-RCR design approach is generally most appropriate for these and other large equipment units or devices.

LCC design under uncertain conditions. One of the few certainties about the business environment is that it is always changing. Under uncertain conditions, equipment designers are expected to forge ahead with their work, despite a lack of basic parameters — such as planned production output, maximum equipment investment cost, and product costs (or market prices) — that are usually regarded as prerequisites to equipment design.

Taking an equipment investment figure from a set of vague estimates to establish the prerequisite data for equipment investment not only risks major financial losses but also puts the future of the company at some risk. The best course of action is to determine which of the fluctuating uncertain elements poses the least risk and incurs the least investment cost in terms of LCC. The specific method for doing this is described in detail in Chapter 4, as are the other three types of LCC design (minimum IC, minimum RC, and IC-RC reduction).

Design for Flexibility

Design for flexibility means designing equipment that can accommodate changes in the production system made in response to fluctuations in the business environment without requiring investment in new or additional equipment. There are three main approaches to design for flexibility.

1. Variable production yields. After equipment designers have made production yield estimates based on market fluctuation predictions, the market is not always kind enough to fluctuate in the manner predicted. Product sales usually fluctuate from month to month. Often, over the course of a product's life cycle, sales are brisk during the first few months but then slack off as competing products start gaining ground. Under such changing conditions, factory equipment must be built for flexible output volumes to avoid costs incurred by losses, additional investment, or unnecessary investment during the period of peak production yield.

2. Diverse product models. The trend toward diverse production models is expected to grow more pronounced in the future. On specialized production lines, short product lives make it impossible to recover equipment investment costs. To avoid this problem, look for equipment that can produce a variety of models (mixed production); in other words, design for flexibility aimed at product diversification. Flexible manufacturing system (FMS) has become a standard term in manufacturing industries referring to design for flexibility combined with factory automation design.

3. Successive models. In almost every industry, new models must replace old models regularly if the company is to keep pace with changing market needs and remain competitive. Products change as market needs change. For example, in the Japanese passenger car industry, minor changes are made in each car model every year, and, on the average, a full model change is made every four years. Production lines are always modified to meet new car model specifications.

Design for flexibility for successive models means keeping these types of changes in mind when designing equipment so that the equipment can accommodate a wide range of product specifications from the start without requiring extensive modification or expansion. Equipment designers must anticipate the types of model changes that may be made and should design factory equipment so that only minor equipment parts need changing to retool for a different model at minimal expense. Although this is difficult to do for some types of products, it will allow substantial cost savings whenever possible.

In all three approaches to design for flexibility, it is rarely easy to build flexibility into equipment hardware. It cannot be done by merely imitating other companies or manufacturing sectors. A designer must combine a basic approach to design for flexibility with a firm grasp of the product's characteristics and intelligent application of available production technology and equipment development resources. (See Chapter 5 for additional discussion of the three approaches to design for flexibility.)

Low-cost Automation Design

The trend toward automation and workerless factories is an inevitable response to the need for increasingly higher product quality and lower production costs. Automated equipment is often plagued, however, by the dual problems of enormous investment costs and frequent product defects. These two problems make it difficult to reach the break-even point for profitability within a short time.

The answer is development of highly reliable, low-cost automated equipment. In the areas where such equipment is needed most, namely, assembly and inspection processes, strict quality assurance and low initial costs are points that must be emphasized. While assembly and inspection processes generally can be automated using conventional design approaches, it is often impossible to achieve both high reliability and low cost this way. A detailed description of automation design is beyond the scope of this book. The approach briefly proposed here, however, aims at low-cost automation (high reliability plus low-cost fabrication) via the following methods.

1. Before drafting concepts for the equipment, analyze how the task is done manually, eliminate all unnecessary functions, then determine which equipment design plan will provide the simplest equipment configuration at the lowest cost.
2. Distinguish between work that people do better (intellectual or organizational work) and work that machines do better (simple operations).
3. When determining methods for building required operational functions into the equipment:
   - Mechanize operations that machines do better.
   - Replace operations that people do better with operations better suited to machines by applying mechanical motion design (MMD) to the product design, the materials used, or the equipment fabrication method.
4. Once you have determined the mechanical function design (marginal design), start on the concrete hardware (equipment) design. While doing this, be sure to evaluate reliability and the economics of adding new devices to provide more completely the required functions.

Intrinsic Safety Design

The more automated factory equipment becomes, the more dangerous it is to troubleshoot or maintain. Intrinsic safety design aims to prevent completely accidents caused by factory equipment. It is not enough simply to meet the safety standards and maintain safety design standards based on past experience.

One goal of safety design is to create equipment that can be stopped from a safe distance when malfunctions occur. Another is to guarantee that the machine will stop automatically whenever someone approaches it, whether it is functioning correctly or not. In other words, the machine should already be stopped and in a safe condition whenever anyone approaches it.

Accidents involving automated equipment. Automated equipment is more prone to breakdowns during operation because it handles materials, parts, or products at high speeds and uses precise and complex mechanisms to perform processing or assembly tasks. When malfunctions occur, the operator may try to repair the automated machine while it is operating. Since automated equipment generally includes delicate computer chips, sensors, or other electronic devices, it must be built for high reliability to help reduce complicated maintenance and fine-tuning.

Nevertheless, recent labor statistics show that accidents related to automated equipment are on the rise. Such accidents can stem from several types of causes:

- Human error in equipment operation
- Worker entering the danger zone to fix a malfunction manually
- Worker entering the danger zone while the equipment is in temporary pause status
- Equipment malfunction due to a faulty control device or sensor
- Equipment malfunction due to unforeseen interlock failure between subordinate device and main device

It is ironic that workers can be hurt by machines intended to create a "workerless factory." The ultimate source of this problem lies in the design methods used to create the automated equipment in the first place. The best conventional design methods for automated equipment not only meet legal regulations and experience-based equipment standards but also incorporate safety data and advice from the factory to further enhance safety assurance.

What is intrinsic safety design? Even the best conventional design methods are not enough to eliminate completely the causes of accidents involving automated equipment. The missing ingredient is intrinsic safety design, which is founded on the premise that human errors lead to equipment breakdowns defined as follows:

- The equipment should ensure safety by covering for any errors or omissions that can be made by its human operator.
- The equipment should be able to set itself into a safe condition when a failure in its oil, air, or electrical supply causes a malfunction to occur.
- The methods, speed, and positions used by the equipment operator must remain within safe limits at all times.

Equipment designed for intrinsic safety might include a device that will detect malfunctions or abnormal conditions before an accident occurs. If it cannot correct the problem itself, the equipment will at least move into a safe position, automatically stop itself, and sound an alarm to alert the human operator to the problem. It might also include a device that tells the operator which part is malfunctioning. Other intrinsic safety design measures include safety-check control functions that can detect errors by human operators or that can prevent the equipment from endangering the operator.

These intrinsic safety design methods should be applied in the early stages of equipment design. Consequently, equipment designers should start by listening carefully to what factory operators have to say about a machine's design. This is an important step not only for intrinsic safety design but for equipment development in general. Unfortunately, designers of automated equipment tend to emphasize functions and performance only and overlook the risks that certain parts of the equipment may pose for operators on the false assumption that the equipment will not be handled by humans.

Technical approach to intrinsic safety. The technical approach to intrinsic safety begins with the elimination of as many potentially hazardous aspects of the equipment as possible. If any such hazards are too essential to the machine's operation to eliminate, separate them completely from the operator's sphere of work and add foolproofing devices to prevent human contact with all hazardous areas or parts of the equipment. Even that will not eliminate all risk of accidents, so consider failsafe control techniques, such as providing all necessary safety protection devices for human operators to wear when they are near the automated equipment. In other words, it is better to address the intrinsic safety issue not only by making the equipment safe but also by providing defensive devices for human operators. Instead of merely reducing the risk of accidents, you can ensure accident-free operation by equipping both the machines and the operators with carefully designed safety devices.

Designer's attitude toward safety. No matter which approach is taken, intrinsic safety design will not succeed unless the equipment designers address potential hazards from every conceivable angle. Merely recognizing all logical risks does not eliminate them. Unfortunately, it is human nature to occasionally discount real risks as once-in-a-blue-moon occurrences and to confuse primary problems with secondary ones. You cannot make up for lack of intrinsic safety by merely adding alarms to equipment, for example, or by providing protective clothing for workers. There is simply no substitute for building safety into the equipment from the start.

Built-in safety measures include sequence controllers that use an interlock function, machine controllers that operate independent devices with only localized effects on the equipment, cutoff or escape-route devices that separate workers from machines, and other devices suited to particular equipment characteristics or environmental conditions. More and more equipment includes computers or other system control devices, but people rely too heavily on their relays or limit switches to protect them from danger when a malfunction occurs. Difficult as it sounds, equipment should be designed to be inherently safe even when its safety assurance devices malfunction.

At the earliest conceptual stage, the person in charge of the equipment design department should form an equipment safety inspection team with relevant persons in other departments. This team should perform preliminary evaluations of equipment safety at each step in the design and development process.

Such watertight safety assurance activities are among the efforts you must make as part of any truly effective early equipment management program. The basic approaches to intrinsic safety design described above can be summarized as follows:

- Mistake-proof design (poka-yoke design to catch or prevent operator errors and other abnormalities)
- Failsafe design (designed to return to safe position)
- Nonmechanical design (designed for fewer moving parts)
- Derating design (designed for higher maximum rated load)
- Independent overlapping systems design (several independently operating devices)
- Redundant design (several similar devices, effectively spaced)

Improved Production Technology and Fabrication Methods

The specific technologies relating to production technology development are beyond the scope of this book; however, it is through a company's technological strength that the various issues just described are resolved. In coming years, greater demands will be placed on early equipment management. To meet those needs, equipment developers must maintain a firm grasp of current market and factory needs and technological advances while working daily to make progress in the company's own R&D. The following are just a few of the R&D needs.

- New materials applications, such as new ceramics and metals
- Cutting-edge applications of fine mechanical technologies, such as LSIs and laser processing
- Expanded computer applications, such as CAD, CAM, AI (expert systems), and CIM
- Automation technologies and related self-diagnosis technologies
- Measurement and control technologies for higher reliability

Carrying Out Early Equipment Management

The procedure for carrying out early equipment management is described in detail in Chapter 2, so it is outlined only briefly here.

From Product to Equipment Development and Design

For optimal results from early equipment management, you should determine the fabrication methods and equipment format early — when modifications can still be made easily at the development and design stages. List goals and gather ideas for factory-friendly products; consider and include new fabrication methods in product designs to foster development and design of more appealing products.

The goals of early determination and innovative application of fabrication methods are more easily achieved if you pursue them at the following three points, beginning with the earliest development stage:

1. Analyze needs reflected in the production of current products thoroughly before launching into the development and design of new products.
2. Continue to focus on fabrication and equipment issues after commencing new product development.
3. Do not request equipment design modification after the prototype or test evaluation stage when new product development is nearing completion.

Establishment of Equipment Design Mission

Before actually designing fabrication methods and equipment, equipment designers must set clear quantitative goals, then draft and adopt a specific plan for achieving those goals. In other words, the work of designing equipment becomes the work of achieving a mission with specific goals. The main themes in this mission are methods for developing process quality and for controlling costs (in terms of LCC). After setting target values for quality and cost, you must inevitably overcome many hurdles to reach those targets. The key is to start with a clear idea of which technical issues are involved. Goal-setting is meaningless unless it is accompanied by down-to-earth planning.

Preliminary Evaluation

Now is the time to begin a technical study of the fabrication methods and equipment specifications that will be needed to achieve the design mission. Mistakes or insufficient study at this point will have an enormous negative impact later. About 80 percent of the life cycle costs should be predictable at this stage.

To achieve the design mission, make a careful evaluation (design review) of the fabrication methods and equipment specifications decided upon thus far, identify any apparent problems, and revise the fabrication methods and equipment specifications if necessary. The best way to do this is by breaking down the preliminary evaluation process into two stages: one each for fabrication methods and equipment specifications.

Step-by-Step Management

Once you have come up with a working set of fabrication methods and equipment specifications and the beginnings of actual designs, you cannot afford to lose control of early equipment management. Carry out careful debugging at every step from review of fabrication methods and equipment specifications to initial operation of the equipment. This process is called step-by-step management.

Collecting and Using Maintenance Prevention (MP) Data

When developing and designing new equipment, never overlook the need to collect and use data concerning previous equipment designs. Study the improvements suggested for that equipment and try, if possible, to incorporate all essential improvements into the new designs. This feedback process can be difficult since equipment operators and maintenance staff have little or no contact with equipment designers from day to day, Therefore, make a point of educating both groups of people about the need for closer cooperation in improving the company's technological strength; set up procedures for gathering MP data and working them into new equipment designs and standards.

Production Setup Process and Daily Management

Production setup generally refers to the process of getting new products into the mass-production stage. Typically, new equipment is most conveniently installed during this period, except when the production setup involves only revisions for a new process (a new manufacturing method or fabrication method) or other simple changes. The most important tasks during production setup — those requiring much time and labor — are the equipment design-related tasks that begin with process planning and end with early mass production.

Two themes dominate each phase of early equipment management: the design of better equipment and strict adherence to the schedule for on-time product shipment. Unless you accomplish production setup tasks systematically, it will be difficult to do justice to both themes. This is why you must draw up a clearly defined procedure for production setup. While also leveling design labor hours, plan detailed setup tasks that are both thorough and efficient; then establish and adhere closely to a daily schedule (for design, materials procurement, and mass-production startup). Calculate the daily schedule backward from the product shipment deadline date to ensure prompt delivery.

The Engineer's Two Pitfalls: Equipment Ergonomics and Cost Efficiency

All equipment design engineers must watch for two pitfalls as they work their way through the problems encountered during the early equipment management process: misunderstood or overlooked ergonomic and economic issues that surface during the equipment development/design stage. The following sections outline a fundamental approach for avoiding these two pitfalls that is also valid for other aspects of equipment design discussed throughout this book.

Equipment Design and System Orientation (Ergonomics)

Designing the visible equipment (the hardware) is not the chief concern of equipment design during early equipment management. Launching prematurely into actual equipment design only produces equipment with built-in high cost and low reliability.

What is system orientation? In this context, system orientation means designing each process as if it were a system of its own. Figure 1-4 outlines a simple process. In the figure:

- *I* refers to the materials being input to the process.
- *O* refers to the processed goods that are output from the process.
- The process includes a function that converts the input *(I)* into output *(O)*. This function consists of operators *(Mn)*, machines *(Mc)*, and measurement *(Mt)* that link the input and output.

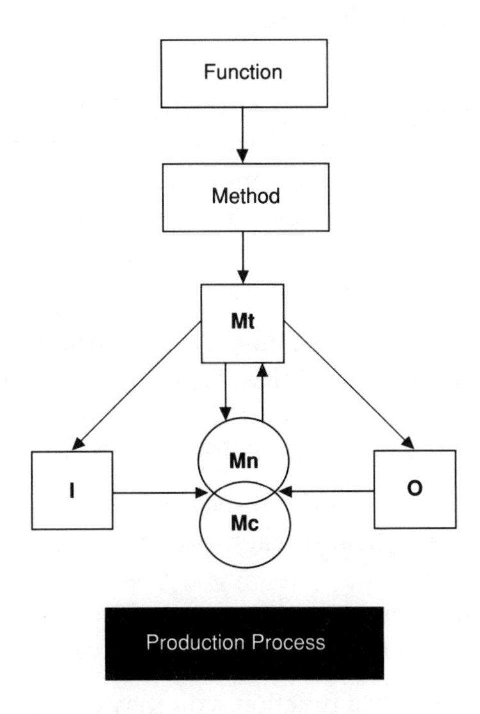

Figure 1-4. System Model of Production Process

Measurement (*Mt*) quantifies the various elements that make up the process and also controls the interrelated statuses of these elements (for both manual and automatic processes).

The process system must meet the following four requirements. Please refer to the system model shown in Figure 1- 4 while reviewing these requirements.

1. ***Object.*** The object (goal) of this process is to produce the output *(O)*. For example, you must quantify the output's processing quality standards and the time-based production output.

2. ***Function.*** The function of this process is to convert the initial characteristics of each workpiece as input *(I)* into its processed characteristics as output.

3. ***Elements.*** The five elements in this process are input *(I)*, output *(O)*, operators *(Mn)*, machines *(Mc)*, and measurement *(Mt)*.

4. ***Relationships (among elements).*** Relationships among the elements may include measurement *(Mt)* data and actions based on those data; operations performed by operators or machines *(Mn/Mc)* to convert input into output $(I \rightarrow O)$ and interactions between operators and machines *(Mn+Mc)*, such as operations and maintenance.

A process can be seen as a system model, as shown in Figure 1-4, but bear in mind that design work consists of fulfilling the system's four requirements as cheaply as possible. If we think of design as simplification, the tasks are to

- meet the product design requirements via process planning by quantifying each process's input and output characteristics,
- establish elements (and relationships among elements) that fulfill the function at minimum cost.

Once you identify a function, you may find several ways of fulfilling it. Therefore, it is best to start with a clearly defined function since the entire design is conceived as building a system that fulfills the function. Next, make sure that the various elements *(I, O, Mt, Mn,* and *Mc)* do not conflict with each other, and that they are designed to interact harmoniously within the overall system. Consider the design itself as one of the elements composing the system. (See Chapter 3 for further description.)

Human-machine systems. In the process system described earlier, the operators and machines *(Mn* and *Mc)* have a direct function (processing) with regard to the material input *(I)*. A typical example is a semiautomated process that includes a combination of human work and machine work. In this case, operators and measurements *Mn* and *Mt* support the function of machines *Mc*.

The role of the operators is to carry out correctly the function of the process. Measurement data may support the function of the machine automatically or as a result of the action of the operator. The role of the operator is also to support the function of measurement.

Thus, the main elements in this process are operators and machines *(Mn* and *Mc)*. It is called a human-machine system because almost all problems related to breakdowns or defects (except for defective material element *I*) in this process are handled by the function provided jointly by the operator and machine elements. Accordingly, it is vital that the two main elements *Mn* and *Mc* have clearly defined functions and are designed to work well together in support of the process function.

In summary, define clearly both the function of the machine and the design that supports this function through measurement (*Mt*); also define the role of operators *(Mn)* in executing the overall function of the process. Finally, ensure that these functions and roles involve compatible equipment characteristics (functions done best by machines) and human characteristics (functions done best by people).

A simple example illustrating this point is a process in which a workpiece is spray-coated with glue from a spray gun (Figure 1-5). To ensure the glue-coating quality specifications shown in the figure:

- Set up the function of element *Mc* (the auto glue supply system that includes the gun, pump, and tubing) in terms of the flow rate per time unit.
- Set up the function of element *Mn* (the person who operates the glue gun to coat the workpiece) in terms of squeezing the gun's ON trigger at the specified glue points and maintaining a steady pattern and speed.

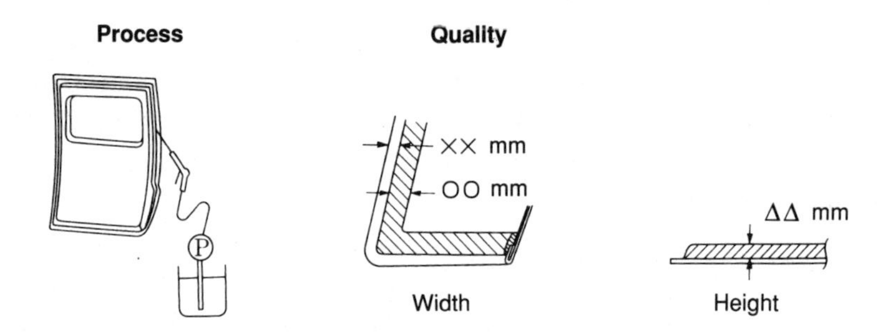

Figure 1-5. Example: Spray-coating Process

Everything ought to proceed smoothly, but it is not as easy as it looks. In fact, this process arrangement contains two fundamental errors.

First, element *Mc*'s function of ensuring a stable flow rate cannot be done reliably unless the temperature of the glue is also controlled (to control the glue's adhesiveness characteristic). Second, maintaining a precisely stable pattern and speed, as called for in the figure, is beyond the capacity of human operators.

Thus, this example is typical because its design fails to ensure accuracy and continuity for the process system's function. Accuracy means that the function does not become impaired over time. For instance, the gun's discharge rate may depend on the temperature of the glue, or the glue pattern may be off course in at least one spot. Continuity of function means that the function does not become impaired over time. For example, a human worker operating the glue gun naturally grows tired over time. His or her fatigue causes fluctuation in the speed of glue application. A worn piston ring in the glue pump would also reduce the glue discharge rate.

Therefore, it is the equipment designer's duty to design a human-machine system with built-in high reliability: a system in which the accuracy and continuity of function can be maintained reliably. (For further discussion, see pages 129-155 in Chapter 3.)

Equipment Design and Economy

No equipment designer can afford to lose sight of the economy principle. Equipment should be designed to help the company maximize profitability. The pursuit of economical equipment design includes the following three strategies:

1. Setting cost targets. Set targets for equipment costs (LCC) based on a suitable profit assurance margin. A company's survival depends on sustaining profitability, in part by conserving cost-related expenditures. Equipment costs are too important to consider only after the equipment design is finished. Instead, cost should be a major influence on the equipment design process from the beginning. The equipment designer's mission includes not only the cost factor, however, but also product quality, production capacity, and scheduling. It is easy for one factor's requirements to conflict with those of another.

Setting cost targets is a challenging task: each target must be neither too ambitious nor too easy, and it must be logical — backed by a plan oriented completely toward reaching the target. Avoid making plans that are beyond technical capacities. The antidote for conflicting factors is compatible goals. Technical ability and accurate support data ensure compatibility.

2. ***Drafting a design plan.*** The first goal-oriented design plan will be more effective when cost parameters are applied later to create a new and better design plan. An economic evaluation method is necessary, however, to clarify which kinds of problems must be solved to lower the equipment costs.

3. ***Selecting the most economical proposal.*** Select the equipment proposal that offers the greatest economy. The more design proposals are available for comparison, the better are your chances of selecting a truly superior one. The time and effort involved in drafting numerous proposals is wasted unless there is a reliable method for identifying which is best.

Accurately Pursuing Economy

There are several ways of evaluating economy in equipment investment, but it is still far too easy to make mistakes in applying these methods. For instance, how can LCC be used to help determine overall cost effectiveness? A simple example should explain this method. Cost effectiveness can be expressed as shown in equation 1-1.

$$\text{Cost effectiveness} = \frac{\text{SE (system effectiveness)}}{\text{LCC}}$$

System effectiveness is the impact of LCC on the overall system's output in terms of the number of design goals achieved, the average annual production yield, availability, and so on. The following table shows SE and LCC values for two proposals (A and B).

	System Effectiveness (SE)	Life Cycle Cost (LCC)
Proposal A	600	300
Proposal B	700	500

We can determine the cost effectiveness (CE) value for each proposal as follows.

$$CE_A = \frac{600}{300} = 2.0$$

$$CE_B = \frac{700}{500} = 1.4$$

The CE values suggest that proposal A is the better proposal, but is this really the case? Look at this comparison of proposals again, this time from the perspective of economic engineering. As can be seen in Figure 1-6, proposal B involves 200 more points as additional investment, while adding only 100 points to the SE value. This means that:

$$CE_{B'} = \frac{700}{500} = 0.5$$

Figure 1-6 shows that both plans have the same 100-point SE value on 200-point investments. However, if the recipient of proposal B's 200 investment points is proposal C (it should not be proposal A because that would cause a 200-point investment surplus, so instead it should go to safety assurance measures presented in proposal C), the SE points would be 40 (CEc = 0.2), which makes proposal B the better investment. This was made clear only by turning the general concept of LCC around and using reverse logic.

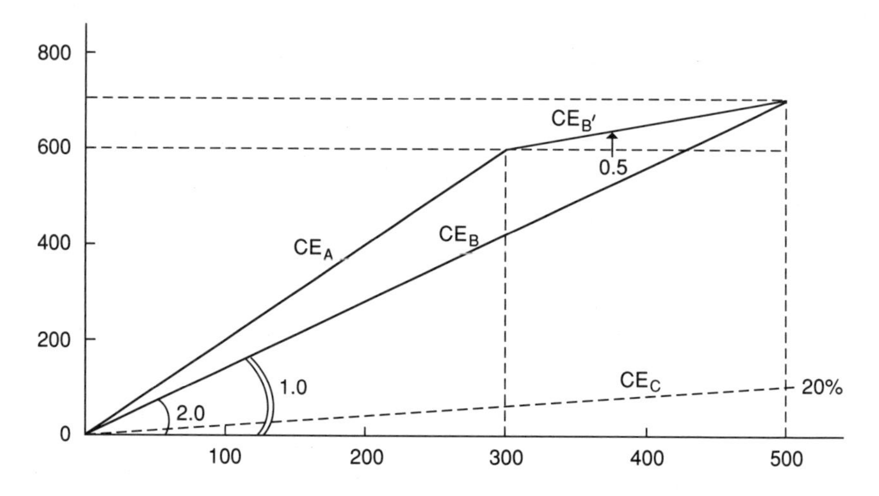

Figure 1-6. Comparison Based on Economic Engineering Principles

Other evaluative criteria, such as the investment recovery period and the return-on-investment ratio, are not criteria for comparing economy. They are merely measures of investment safety.

It is easy to make mistakes in evaluating the economic characteristics of equipment design proposals. It requires accurate data and a solid understanding of the logic and principles behind economic comparison of design proposals.

The most effective method for pursuing and evaluating economy in equipment designs is the economic engineering method described above. This method was devised by Dr. Shizuo Senju, professor emeritus of Keio University in Tokyo,

Table 1-1. Checklist for Evaluating Investment Proposals

A. Check calculation method used for proposed investment plan

1. Are effects of investment (including cash flow) correctly understood?
   - Is the cash flow arrangement appropriate? (Compared to what?)
   - What was the time frame (and area) used for consideration?
2. Was the resulting judgment criterion appropriate?
   - Were the payback period and ROI interpreted correctly?
   - Were current price, annuity value, final value, and additional interest all used appropriately?
   - Were any incorrect measurement approaches used (such as cost per ton, etc.)?
3. Were appropriate analytical methods used?
   - Were there any errors in the use of the annuity value method, final value method, ROI rate method, payback period method, cost, profit, prices, rates, etc.

B. Check on the extent to which sensitivity analysis and risk analysis were used

1. Check on the extent to which uncertain factors were used.
2. Do you know which of the uncertain factors warrants the most caution in use?
3. Do you know the point at which alternative plans become more advantageous?

C. Check on previously considered alternative plans

1. What plans other than the ones presented were studied?
2. Did you look into price margin efficiency?
3. Did you sort out the different intangible factors?
4. Were alternative plans considered?(Unavoidable ends do not necessarily mean unavoidable means. Investments cannot simply be excused as unavoidable.)

D. Were support, cooperation, and guidance given to the people who investigated the investment plans?

Source: Shizuo Senju, "Strategic Management and TP Management," *Production Management Magazine*, October 1985.

Japan. (See Table 1-1.) This economic engineering method for evaluating economy factors in equipment design will be used throughout this book. For further discussion of this method's application in LCC design, see Chapter 4. The basic principles of equipment development and design for TPM summarized in this chapter are discussed in more detail in Chapters 2 through 5.

Sources:

Senju, Shizuo, and Tamio Fushimi. *Fundamentals of Economic Engineering* (Tokyo: Japan Management Association, 1982).

——— . *Applications of Economic Engineering* (Tokyo: Japan Management Association, 1983).

Senju, Shizuo, and Zentaro Nakamura. *Easy Economic Analysis*. (Tokyo: Japan Management Association, 1987).

2

Implementing Early Equipment Management

This chapter addresses the range of activities that constitute early equipment management — initially developing and designing products and equipment, establishing a design mission, conducting the preliminary evaluation and step-by-step control, and collecting and using MP data. The first section explores the relationship between equipment and product development and design.

FROM EQUIPMENT TO PRODUCT DEVELOPMENT AND DESIGN

Obviously, the transition from product development and design to equipment development and design must be handled in a positive way by a company's product development and design department. What kinds of activities must the product development and design department undertake to prepare for the transition?

Factory-friendly Products

A factory-friendly product is, simply, a product that is easy to produce. In other words, it is manufactured under safe conditions through low-cost production methods and simple manual or mechanical operations. Figure 2-1, for example, shows factory-friendly requirements for a product (or product part) produced primarily through cutting. The criteria included easy processing and assembly, easy-to-meet standards, and easy clamping and cutting. Other issues addressed included preventing cutting debris from entering parts, preventing errors and machine-generated defects, and promoting easy automation.

Regarding this last criterion, consider the following items when drafting general requirements for easy-to-automate equipment based on part-specific designs.

- Can this part be eliminated?
- Can this part be integrated with an adjacent part (such as by forging, plastic molding, pressing, or diecasting)?
- Can the part's shape be made simpler?
- Are the plate thicknesses appropriate, or are they too thick?
- Are part strengths consistent?
- Does the design make processing difficult in any way (such as in the thickness of forged parts, the bending radius of pressed parts, the depth or angle of drilled holes, required cutting, part shapes that require nonstandard tools, etc.)?
- Are specifications for dimensions, finishing, and/or shape unnecessarily strict?
- Are the materials appropriate?
- Are the processing methods appropriate, especially when related to production yield?
- Are the surface-processing methods appropriate?

- Is the part's shape conducive to removal and recycling of surplus material?

Five Measures for Factory-friendly Products

Five measures are necessary to establish the kind of factory-friendly product development and design described above. First, gain a detailed understanding of the problems that have made current products less factory-friendly. Incorporate solutions to those problems in the new product design by:

- Collecting and using feedback data about current products.
- Analyzing the production processes used to make current products to discover and resolve process-related obstacles to factory-friendliness.

Next, carry out design reviews for every stage, from product conceptualization to prototype fabrication, testing, and evaluation to resolve problems and work out solutions that can be incorporated into new products. To do this,

- Analyze the production processes planned for making the new products to discover and resolve process-related obstacles to factory-friendliness.
- During design review of the new products, identify risks of defect production and find ways to reduce those risks.
- Find solutions for mass-production problems identified during the prototype, testing, and evaluation stages.

Two Examples of Factory-friendly Product Design

The first example describes a design for an easy-to-assemble product; the second shows how a design that standardizes

Design of factory-friendly compressor → Easy to process

	Basic Requirements	Standard Parts
Easy-to-meet standards	• Assurance of flat surfaces • 3-point flat surfaces • Large 3-point pitch • Avoids slide dies (insert dies)	• Housing = ϕ10 - ϕ12, • cylinder width = approx. 20 cm • 120° intervals • Close to product's outermost points • Either fixed or movable
Easy to clamp	• Flat or curved clamp fingers • Close to reference surface • No divided surfaces • Can clamp at least 3 units • Can be positioned	• Clamp fingers that open at least 5 mm long and 10 mm wide • Must be on same radial as reference plane • No uneven surfaces or burrs • Dimensional variance 0.6 or less • Use of bumps or dents for clamping
Easy to cut	• Few blade replacements	• When diecast is 1 mm or less or when Aℓ cast is 2 mm or less
	• No intermittent cutting	• As much as possible
	• High product hardness	• Wall thickness
	• Shallow holes	• As much as possible
	• Wide holes	• As much as possible
	• Large tolerances	• Circle-grade fit (or better)
	• Avoids slanted holes	• As much as possible
	• Avoids intersecting holes	• As much as possible
	• Attach C face to black (forged) surface	• 45° (if possible, 30 - 60°)
	• Slight deviation in replacement	• 0.5 or less
	• Short internal thread	• Screw strength should be 1.5 - 1.8 times the diameter
	• Large chamfered area at screw edges	• C1
	• Deep base under internal thread	• 2.5 times the screw pitch
	• Large screw diameter	• M5 or larger

	Requirement	Items	Measures
Easy to assemble	Prevent cutting debris from entering	• Shallow holes	• As much as possible
		• Wide holes	• As much as possible
		• Perforated holes	• As much as possible
		• Large tapered holes	• 5 degrees or more
	Easy to design	• Clear indications of reference surfaces (or points)	• Indicated on surfaces, hole centers, and R centers
		• X-Y indications	• As much as possible
		• Indications on actual surfaces	• As much as possible
	Error resistant	• Cannot be assembled incorrectly	• Poka-yoke design
		• Errors are obvious at first glance	• ID coding (by color, shape, markings, etc.)
	Defect resistant	• O-rings are fixed or flat surfaces	• O-ring groove in the bottom part during assembly
		• Fixed O-ring for cylinder surface on the spindle side	• No slack in inner fit (spindle should be slightly wider than O-ring inner diameter)
		• Fewer parts	• As much as possible
		• No selective fitting	• Designed for large tolerance
		• Fittings are chamfered at area receiving fitting	• No chamfering from 30 - 45°
	Easy to automate	• Washers kept together with bolts and nuts	• Makes washers harder to lose
		• Use hex bolts or phillips bolts	• Hex bolts rather than inset-hole bolts, and phillips screws rather than slotted screws

Figure 2-1. Requirements for Factory-friendly Product

parts and changes the fabrication method from spot welding to gluing can facilitate automation.

Case Study 2-1: Locker Cover Design Revision (Ford Motor Company)

In an effort to promote a greater concern for manufacturing needs in its design engineering, Ford formed an advanced group consisting of design engineers and production engineers. This advanced group looked into new manufacturing processes that promised higher quality, and sought to develop designs to accommodate these new processes.

Figure 2-2 shows the design for attaching automobile engine valve covers. This design uses an integrated seal to facilitate automation and achieves a 70 percent reduction in parts, a 52 percent cut in manufacturing costs, and a 20 percent savings in expenditures for materials.

Case Study 2-2: Standardization of Parts and Revision of Fabrication Method for Easier Automation (Hitachi)

At a highly automated production line in Hitachi's Mito Plant (Katsuta, Ibaraki Prefecture), cosmetic panels for standard elevators are produced with almost no human assistance. In this typical example of small-lot, wide-variety production, the cosmetic panels are produced in 335 different sizes and in various colors for each size, which means a nearly endless variety of size-color combinations (see Figure 2-3).

The designers automated the paint color changeover process. They also automated processes centered on changes in panel sizes and, whenever possible, standardized parts such as accessories and braces. This design approach combined a greater variety of products (decorative panels) to suit customer demands with a smaller variety of parts (parts seen by elevator users) for easier and cheaper production.

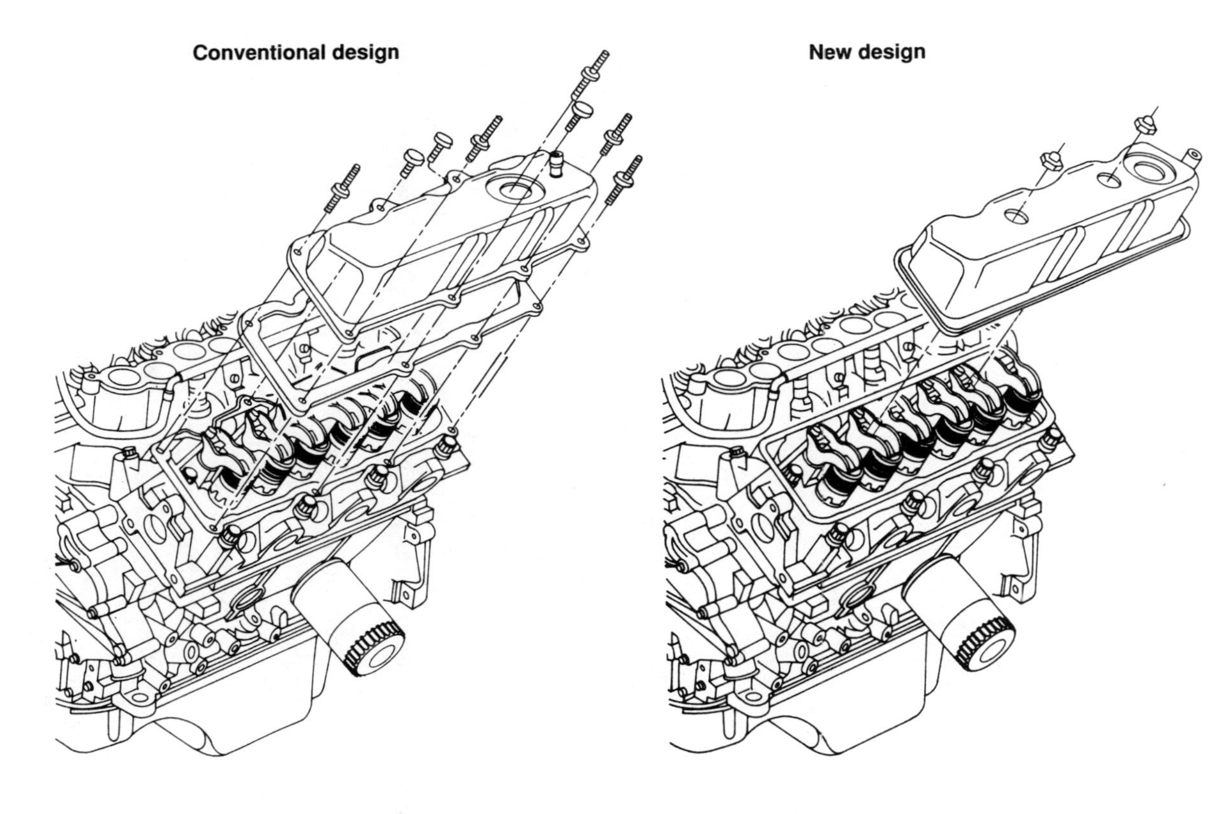

Figure 2-2. Design Revision for Valve Cover

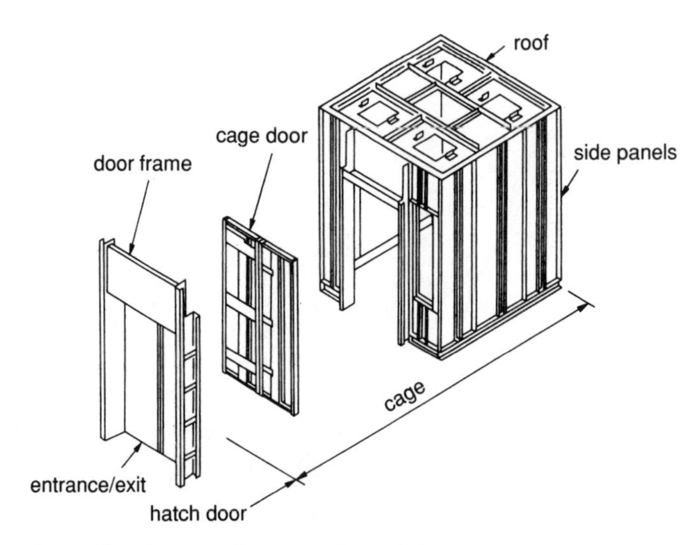

* All manufactured on the same line except for roof and door frame

Figure 2-3. Configuration of Elevator Panels

Hitachi then introduced an automated sheet-metal processing line that combined two previously separate elevator panel processes: bending two panel sides and attaching braces to the other two panel sides. Another improvement attached the braces vertically and closer to the center. As a result, automation became easier to implement and the company achieved a higher return on plant investment capital for the new processing equipment.

Another revision involved a switch from spot welding to gluing as the method for affixing braces. Spot welding tended to produce large warps in the braces, and so the line included a laborious, warp-straightening process before the painting process. Using an adhesive to affix the braces resulted in far fewer warps and may eliminate the warp-straightening process altogether.

The gluing process needed to be designed as an automatic process. The engineers evaluated the adhesive strength of a method in which a double-adhesive compound was mixed on the panel, and decided to adopt this method. Safety studies indicated, however, that the adhesive would not be strong in emergency situations, and that the braces should be designed to receive spot welds at two sites on each side to ensure safety (see Figure 2-4).

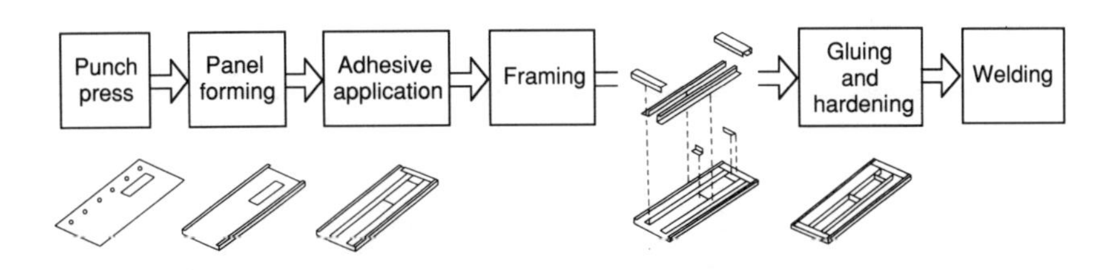

Figure 2-4. Panel Assembly Processes

Table 2-1 shows how engineers sought to make the cutting process easier. A process analysis of the current product identified the needs to be addressed in a similar new product.

FROM EQUIPMENT DEVELOPMENT AND DESIGN TO PRODUCT DEVELOPMENT AND DESIGN

Typically, once the product design plan has come close to its final form, the development process is already moving to the process planning and equipment design stages, so there is little or no leeway for product designers to consider the ideas of equipment designers. As a result, those ideas are not reflected at the prototype testing or test run (design review) stages.

It is difficult to have feedback reflected in product designs when it arrives so late in the process. It is possible, however, to have equipment design input at an earlier stage in the product development and design process, and this input can be effective

Table 2-1. Process Analysis Table for Current Product

Manufacturing process analysis table for current product

Car type: ◯◯ G

Part number: ◯◯◯◯◯ – ◯◯◯

Part name: Socket lower ball joint

Production output: 60,000 units/month

Manufacturing dept: 00

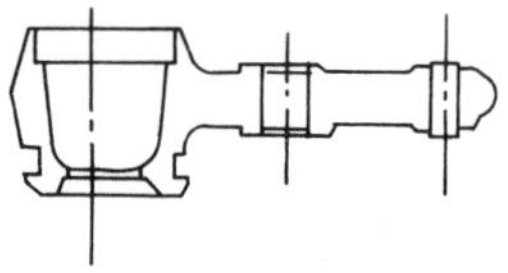

Schematic drawing

Process (machine)	Area to be machined	Defect statistics		
		Defective item	No. of defects	Defect rate
1 Reference surface processing (specialized machine)	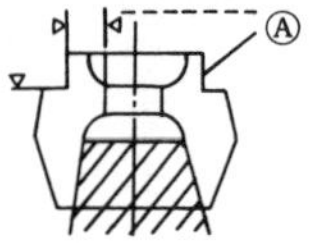 (2 units processed)	a. Off-center	*n* units/month	0.0× %
		b. Depth (–)	*n* units/month	0.0× %
2 Inner diameter and caulking (specialized machine)	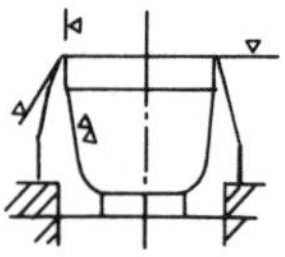 (2 units processed)	a. Damage on inner diameter	*n* units/month	0.0× %
3 Processing for grooves and opening (specialized machines)	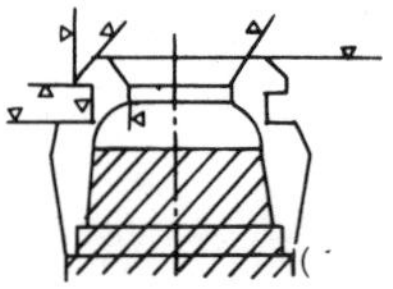 (1 unit processed by 2 machines)	a. Groove diameter (–)	*n* units/month	0.0× %
		b. Through hole (+)	*n* units/month	0.0× %
4 Drilling (specialized machine)	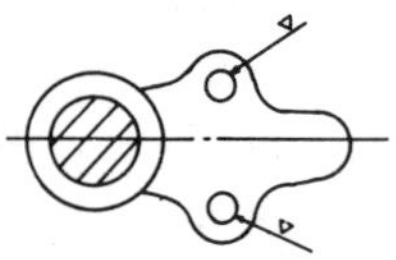 (2 units processed)	a. Uneven hole pitch	*n* units/month	0.0× %
		b. Damaged holes	*n* units/month	0.0× %
5 Screw hole drilling (specialized machine)	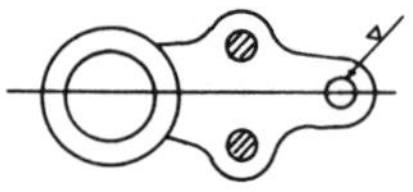 (2 units processed)	a. Uneven hole pitch	*n* units/month	0.0× %

Total weight entered A:	◯◯◯ g	Material expenses		File no.: ◯◯ G.S./-1/2
Raw material weight B:	✕✕✕ g	Material used:	S48C	
Finished product weight C:	△△△ g	Price per kg:	✕✕ /kg	Investigation date:
		Unit price:	✕✕ /unit	March 17, 1987

Yield			Investigators:	Manufacturing dept.: ◯◯
	B/A	70%		
	C/B	75%		
	C/A	53%		

Abnormalities	Causes of defects	Improvement plan
	Error in clamping or attaching	Review shape of clamping jig (production engineers)
	↑	↑
a. Black surface area left on part A	Shaped of forged raw material	Investigate shape of reference surface (production engineers, designers, forge engineers)
	Broken drill bit	Reduce drill bit replacements (designers and forge engineers)
	Occurs during drill bit setup	• Check tool holder and tool precision (production engineers) • Investigate shape and precision of grooves (designers)
	↑	↑
a. Drill-bit service life is too short/too many drill bit replacements (13 or 14 replacements per day)		• Investigate tool materials (production engineers) • Investigate shape and precision of grooves (designers)
	Workpiece setup error	Investigate jig shape and setup method (production and manufacturing engineers)
	Broken drill bit	Check drill-bit material and resharpening frequency (production engineers)
	Workpiece setup error	Investigate jig shape and setup method (production and manufacturing engineers)

in raising product design engineers' responsiveness to equipment designers.

Figure 2-5 includes two charts. One shows the number of proposals sent from the production engineering department to the product design department at different product development stages. The other shows the types of proposals that were made. As can be seen in the charts, proposals concerning new products tend to come at earlier stages, while those for current products are less frequent at the conceptualization and design stages, and more frequent at the prototype and evaluation stages.

Figure 2-6 shows that fewer proposals are made for older current products than for newer current products, and that a much smaller increase is seen in the number of adopted proposals. Figure 2-6 also separates these proposals into different categories. The breakdown for not-yet-adopted proposals shows that proposals expected to substantially reduce equipment expenses were nevertheless abandoned because of constraints on the product structure or design concept, or because of the proposal's tardiness.

Only small improvements, mostly involving relatively minor matters, such as operability, were effected by the adopted proposals. According to the breakdown of proposal contents shown in Figure 2-5, many proposals for new product design concepts and other early design factors dealt with quality or cost and were far more numerous for new products than for current products. Therefore, to get the maximum benefit from early equipment management, work out fabrication methods, equipment concepts, and other key matters at the early stages of product development and design, when there is still some leeway for modification.

Work must begin at the early product development design stages for the following reasons. First, from the perspective of equipment development and design, it becomes difficult to make product modifications once product development has

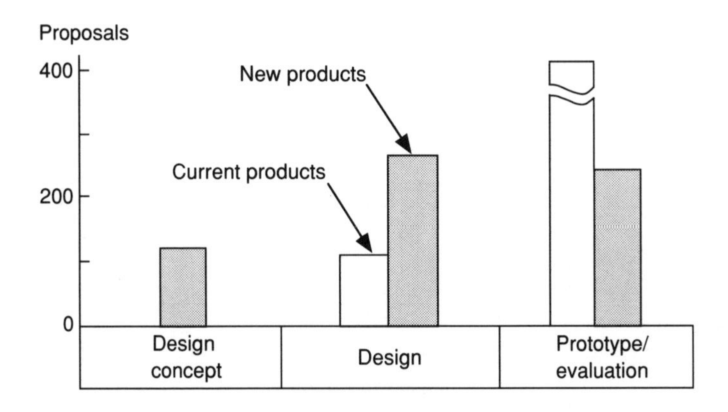

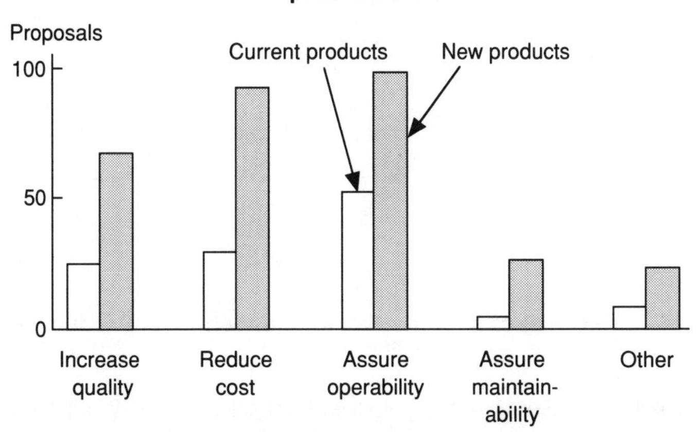

Figure 2-5. Number of Proposals from the Production Department to the Design Department

yielded concrete results. Product development and design must be implemented simultaneously with design of fabrication methods and equipment.

Second, from the standpoint of designing more appealing products, it can be difficult to develop novel products using a design approach based on existing fabrication methods and

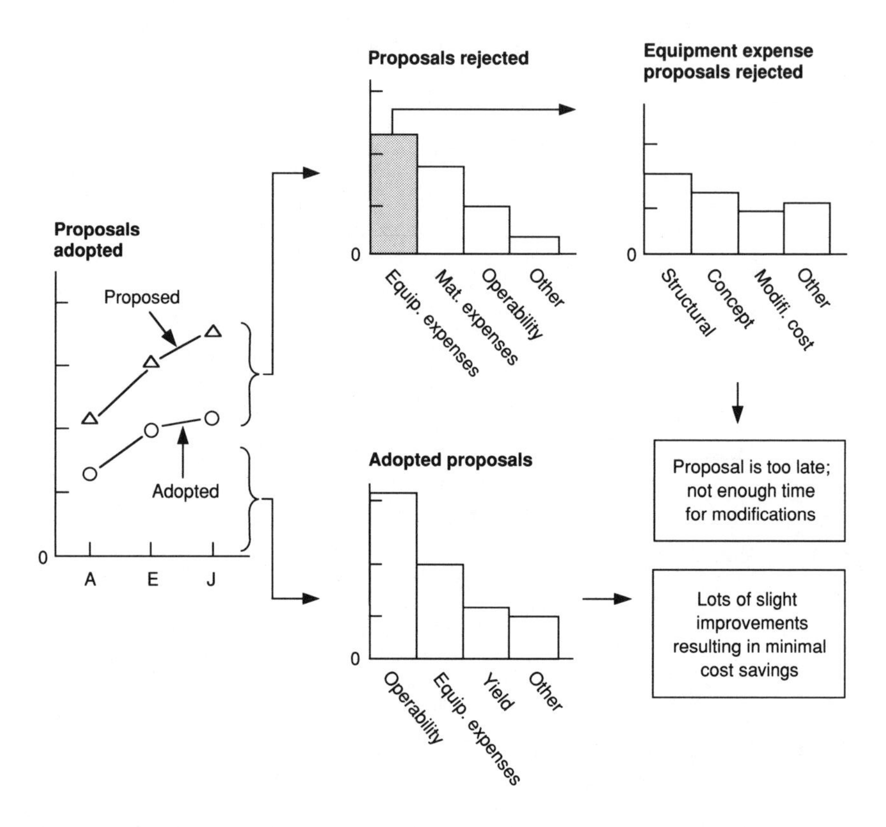

Figure 2-6. Cost Reduction Activities for Conventional Product Design

equipment. Accordingly, product development and design department requests to the equipment development and design department must be sent at an early stage, before the product design has taken firm shape, so there will be enough leeway in the product design process to accommodate any newly developed fabrication methods and equipment plans.

This approach toward product development and design can work at three different stages — in response to current conditions, at the development and design stage, and at the prototype/trial evaluation stage.

Designing New Products Based on Current Conditions

At this stage, you should analyze current conditions that obstruct factory-friendliness and feed the findings back to the product design department as a "product design wish list." In other words, the equipment design department asks the product design department to fulfill as many of the requests on this list as they can while they develop the next generation of products. Designers must attempt this without sacrificing product appeal. If something on the list proves impossible to realize, it is important that the product design department notify the equipment design department of the problem as soon as possible.

Product Design Requests for Economical Fabrication Methods

Generally, not much production engineering-related investigation is carried out at the product conceptualization stage, and so it is not yet clear what kind of fabrication methods will be used. It is often simply assumed that the current fabrication methods will be used for the design in progress. But as products become increasingly sophisticated and complex, they also become more difficult and costly to produce using current equipment.

At the product conceptualization stage, the equipment design department still has the leeway to look at the various fabrication methods and pick those that seem most reliable and economical. Consequently, it is important that equipment designers select fabrication methods and investigate reliability and cost factors at this early stage. Changing the product specifications for a new generation of products presents an opportunity for selecting a more appropriate fabrication method, which, in turn, can result in substantial production cost savings.

Providing New Fabrication Methods to Ensure Product Design Flexibility

The two most important issues for the product development department are market needs and the development of products with strong market appeal. To support the product development department's efforts, equipment designers should not impose constraining fabrication methods on the product developers. Instead, they should work to develop new fabrication methods based on production engineering that preserves as much flexibility in product development as possible. After all, the company cannot survive if the factory-friendly fabrication methods that its production engineers provide work only for product designs that lack market appeal.

The following example illustrates how production engineers can develop fabrication methods that provide effective processing without harming the market appeal of the product design.

Case Study 2-3: Working to Create New Fabrication Methods (Panel Conveyance Method)

When Daihatsu Motor Company made a full model change for its Charade model, the goal was a smooth, integrated look for the car exterior. One challenge in achieving that look was to make the door panels fit in smoothly as single pieces of metal. The engineers pointed out how troublesome it would be to manufacture such door panels: cutting each panel out of one piece of metal would waste the metal cut out to make the window.

The production division issuing the request checked the problems noted by the production engineers. The biggest difficulty concerned the conveyance method for the continuous press process. The method in question, shown in part A of Figure 2-7, consists of a conveyor that brings door panel workpieces to a press operator, who sets each workpiece onto the press. After the pressing is done, an iron hand automatically reaches down, picks up the pressed door panel, and sets it on another conveyor that takes it to the next press process. The door panels are made of high-tensile steel that is thin (0.55 mm to 0.6 mm in width). This method puts a lot of stress on the bottom edge of the panel, a weak point that bends easily.

In response to this problem, the production engineers developed a loader/unloader set for this press process, shown in part B of Figure 2-7. This new method reduces the lowering distance for setting down the door handle during conveyance, thereby lessening the impact on the door handle material. This also keeps the window frame section of the material from being bent. Furthermore, when the door panel material was hand-lifted onto the press under the old method, the slightly different way in which the operator set down the material each time produced some variation in the product dimensions and sometimes led to bent window frames. When the automated loader and unloader were introduced, there was no longer any manual imprecision and the variance problem was eliminated.

Product Design Modification Request (at Prototype/Trial Evaluation Stage)

As already noted, this is a late stage at which to make product design modification requests, but a full-fledged design review at this point can still point out and correct small problems with operability, equipment parts, or other minor matters. This is also the best time to work out mass-production requirements through a prototype test run and evaluation, thereby preventing problems later when mass production begins.

Accordingly, at this stage you should consider the following ways to facilitate prototype testing and evaluation and to predict mass production-related problems:

- Identify additional mass-production requirements.
- Develop new evaluation methods to identify hidden problems.

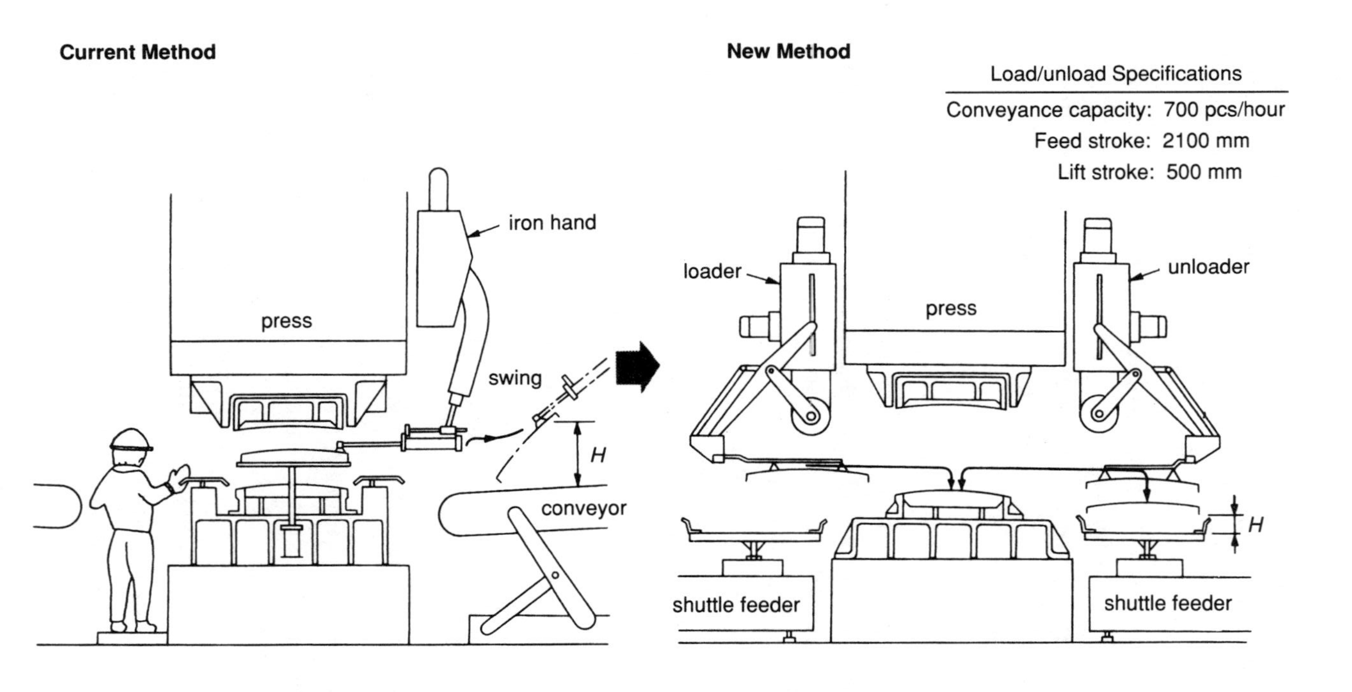

Figure 2-7. Current and New Door Panel Conveyance Method (Press Process)

- Make greater use of standardized materials (such as checklists) for design review.
- Make greater use of highly skilled, highly trained staff and promote systematic responses.

In other words, mass production problems should be discovered *before* the mass-production test run, during the previous stage of prototype testing and evaluation. You must learn to test product functions in the prototype and, at the same time, test for mass-production problems. This allows you to eliminate the mass-production test run stage — which will shorten development lead time — and address mass-production problems at an earlier, more flexible stage of product development.

ESTABLISHING AN EQUIPMENT DESIGN MISSION

It is not unusual for equipment designers to be handed an initial list of design-related problems to be solved. The design must provide answers to several key questions. What kind of equipment is needed? What target values must be met, and what measures must be taken to meet them?

Equipment design is a problem-solving process. Once you have clarified design targets, there is a gap between the initial design plan (including any baseline alternative plans for other conditions or similar process plans) and the target values. The next step is to devise several specific measures to close that gap. The equipment design must therefore include several important functions that support such measures.

So, what is the equipment designer's mission? It is spelled out as soon as you decide the time period, target values, and measures (including technical issues) for the equipment design. In general, the equipment designer must accomplish the following three main tasks:

- Establish process quality and sort out technical issues to provide the quality characteristics needed in the final product.
- Establish an equipment investment estimate and manufacturing cost margin, and sort out cost-related technical issues to meet the planned cost of the final product.
- Establish equipment capacity and sort out technical issues to provide the planned production output.

If equipment designers perform these three key tasks with only vague ideas about target values and technical issues, they may be able to come up with some kind of final product. This lackadaisical approach will result in higher costs and additional problems along the way, however. Moreover, it will not give designers the challenge they need to improve their engineering skills.

Approach to Establishing an Equipment Design Mission

As mentioned above, the three main tasks in the equipment designer's mission concern process quality, costs, and production capacity. Table 2-2 outlines these three tasks. (The tasks do not have to be organized according to the table's format.)

Processes. Process planning (the sequence of processes) must be set up to establish production that supports the product design.

Process quality. Equipment is designed for each of the planned processes, and the quality that is built in at each process must culminate in the desired quality characteristics for the final product. Accordingly, one absolute requirement is that the equipment be designed to provide the required process quality reliably. Specific means for establishing process quality in the equipment design generally include quality deployment, setting process quality standards, setting tolerances, and so on.

Table 2-2. Outline of Equipment Design Mission

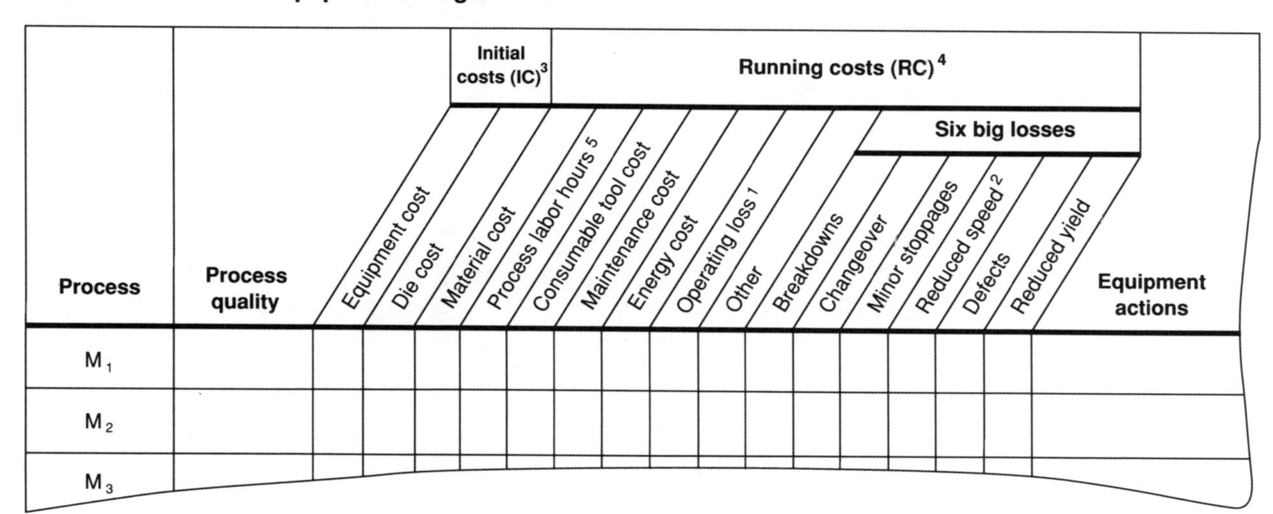

Process	Process quality	Initial costs (IC)[3]		Running costs (RC)[4]														Equipment actions
										Six big losses								
		Equipment cost	Die cost	Material cost	Process labor hours [5]	Consumable tool cost	Maintenance cost	Energy cost	Operating loss [1]	Other	Breakdowns	Changeover	Minor stoppages	Reduced speed [2]	Defects	Reduced yield		
M_1																		
M_2																		
M_3																		

[1] Operating loss is equal to the total financial value of the six big losses

[2] Reduced speed can be expressed in terms of the processing capacity (units per hour), the cycle time, or other measures. For the sake of more flexible capacity, establish minimum and maximum speed values

[3] Initial costs (IC) include all research and development-related costs

[4] Running cost (RC) categories vary according to the equipment's characteristics

[5] Processing labor hours should be expressed as labor hours and labor cost. Sometimes, indirect labor costs, such as those for inspection or conveyance, are large enough to warrant inclusion

Initial cost (IC) and running cost (RC). Working within the range of allowable initial costs (from planning to overall equipment investment costs), assign some of these costs as process-specific equipment investment costs. Working within the range of allowable running costs (the initially planned manufacturing costs), assign these costs as process-specific manufacturing costs. Running costs include fixed costs, such as processing labor hours, and proportional costs, such as material costs.

Six big losses. The six major equipment-related losses include breakdowns, setup and adjustment, idling and minor stoppages, process defects, rework, and startup losses — all factors that lead to higher running costs. From the equipment design standpoint, each of these losses can be converted to a quantified cost value. It is often easiest when designing the equipment, however, to look at the six big losses in terms of their impact on the equipment capacity utilization rate. For instance, breakdowns cause downtime, which can be expressed as a certain amount of time or labor hours per day that can be subtracted from the equipment capacity utilization rate. The expression of production capacity should include a target value (or allowable range of values) for speed (cycle time).

Table 2-3 shows how current conditions are analyzed to find the effects of the six big losses on similar parts or processes when the development of a new product goes from process design to equipment design. This analysis indicates what processes and equipment will be needed. The results of this analysis can lead to more detailed engineering studies to set target values for each of the six big losses while remaining within the overall cost target.

Acknowledging technical issues in the design. After setting target values for the items described above, you need to recognize and organize the related technical issues. A target value unsupported by measures to reach that target is, in effect, the same as no target value at all. When setting targets and working

out the related technical issues, it makes sense to use data from a previous model of the same product as well as data from similar products, processes, fabrication methods, and equipment, as reference sources.

PRELIMINARY EVALUATION (DESIGN REVIEW)

Once the mission and measures for fulfilling it are established, the next steps are drawing up the equipment design and finalizing the specification. Naturally, more ambitious design missions are more difficult to fulfill. Quality and cost factors often conflict directly, but equipment designers must design equipment that solves the problems created by conflicting factors.

The preliminary evaluation is not a matter of judging whether the design is good or bad. Instead, the results of the evaluation guide the development of the design mission by giving a clearer picture of its direction and measures. Consequently, any major revisions of the detailed design for the equipment will incur a substantial loss of time and labor. The time and effort originally invested are gone forever.

Two Steps in the Preliminary Evaluation

Preliminary evaluations must be made at an early stage of equipment design, when there is still a considerable degree of freedom for modification. They should include the following two steps, each oriented toward a particular goal: determining (1) fabrication methods and (2) equipment specifications.

Determining Fabrication Methods

Before deciding on the equipment specifications, determine the fabrication methods. The currently used fabrication methods are not necessarily the most appropriate ones. For example, methods for combining two pieces of sheet metal include welding, gluing, riveting, and fastening bolts and nuts. Within the welding category alone, there are varied methods, such as spot welding, arc welding, and laser welding. A decision about a fabrication method determines approximately 80 percent of total costs. Therefore, cautiously carry out the following two types of preliminary evaluations.

- Select fabrication methods from the menu of available fabrication methods.
- Decide which fabrication method best fulfills the design mission.

Determining Equipment Specifications

Once you select a fabrication method, you may determine the basic equipment specifications, or at least outline those specifications. (So far there should be only a rough design plan.) After determining which level of specifications to generate, evaluate whether or not the detailed functions provided for in the design (down to the components) can fulfill the design mission. Any problems or concerns apparent in the specifications must be resolved before you can arrive at a full set of specifications. Care and thoroughness at this point will prevent problems during the test run phase, and thus forestall several days of added work and added costs beyond the target value. Evaluate whether or not the specifications

- provide for 100 percent of the required quality,
- remain within the target cost value.

These evaluation methods will be described in greater detail in Chapter 3, which covers technical issues in equipment development and design.

STEP-BY-STEP MANAGEMENT

One purpose of early equipment management is to take steps at the various development stages — from planning to installation and test run — to prevent problems from occurring after the equipment has been put to work. Accordingly, the most important part of early equipment management is using checklists at each step (from planning to design, drawing checks, fabrication, witnessed tests, test run, and commissioning). Checking the design or the equipment itself against the checklist and also checking certain equipment model-specific items that may not be on the checklist provides effective debugging at each step, which is why it is called step-by-step management.

Negligence in step-by-step management is usually the underlying factor when problems linger after the equipment is commissioned. In view of this, keep in mind the following five goals of step-by-step management:

1. When determining fabrication methods and equipment specifications, clarify and resolve all uncertainties and concerns that exist at each step.
2. At later stages, when drawing up the design and fabricating the equipment, other (progressively smaller) problems and concerns will arise. Confront these issues immediately and resolve them completely.
3. At each step, check to ensure that any of the problems apparently resolved at earlier steps have been completely eradicated.
4. Keep an accurate record of all activities at each step to monitor their effects on the equipment at subsequent steps.
5. Use different sets of checklists for each step, and note the different steps clearly in the activity log.

Establishing a Project Team

Organizations tend to leave equipment design completely up to the production engineering and equipment design departments. One of the best ways to minimize equipment problems after the commissioning stage, however, is to include equipment operators and maintenance staff in the equipment development and design process right from the planning stage. They can bring different and valuable perspectives to this process and, in turn, they can learn a lot from it too.

Any equipment development and design project should be staffed by a project team that includes, at a minimum, equipment designers and representatives of the factory operations and maintenance departments who take an active part in early equipment management. Indeed, when the project team includes the specialized perspectives and experiences of the equipment operators (in quality, productivity, operability, environmental safety, etc.) and the equipment maintenance staff (in reliability, maintainability, energy savings, etc.), it can be much more specific and effective in identifying or predicting problems and in resolving those problems.

Using Preliminary Evaluation Charts

To do an effective, step-by-step job of early equipment management, the project team needs the right tools. One such tool is the step-by-step preliminary evaluation chart shown in Figure 2-8. This chart can be used to discern problems at each step, work out problem-solving measures, and check the results of those measures.

1. Planning stage. At the planning stage, the project team members (from the design, production engineering, maintenance, and operations departments) discuss various experiences with existing equipment models that are similar to the model to be designed, as well as specifications and other points that require preliminary consideration. As they discuss these things, they fill in the rows in the step-by-step preliminary evaluation chart under the appropriate topic headings (see Figure 2-8). These data will be useful during the follow-up checking at the design and drawing check stage. The project team members should also fill in the column indicating the responsible department and the "Actual" column (in effect, things that must be done).

2. ***Design and drawing check stage.*** The project team members meet again, this time to discuss and implement the items entered in the "Actual" column during the planning stage. If their implementation is successful, they should enter a circle in the evaluation column; if it fails, they should enter a triangle. Next, the team needs a fresh preliminary evaluation chart to fill in the problems column with items that received triangles in the last chart, plus any newly discovered problems. This will be used during the follow-up checks at the fabrication and witnessed test stage.

3. ***Fabrication and witnessed test stage.*** The project team meets to follow up on resolved and unresolved issues from previous stages, as well as to estimate and address potential future problems. By taking more time at earlier stages, considerable time is saved in startup. Through a step-by-step procedure, the team identifies problems before they are built into equipment. Moreover, people grow in skill and understanding by working together in cross-functional teams. Figure 2-9 shows an example of results achieved through the approach.

Application of Step-by-Step Preliminary Evaluation Chart

This section presents an example of how the step-by-step preliminary evaluation chart is used.

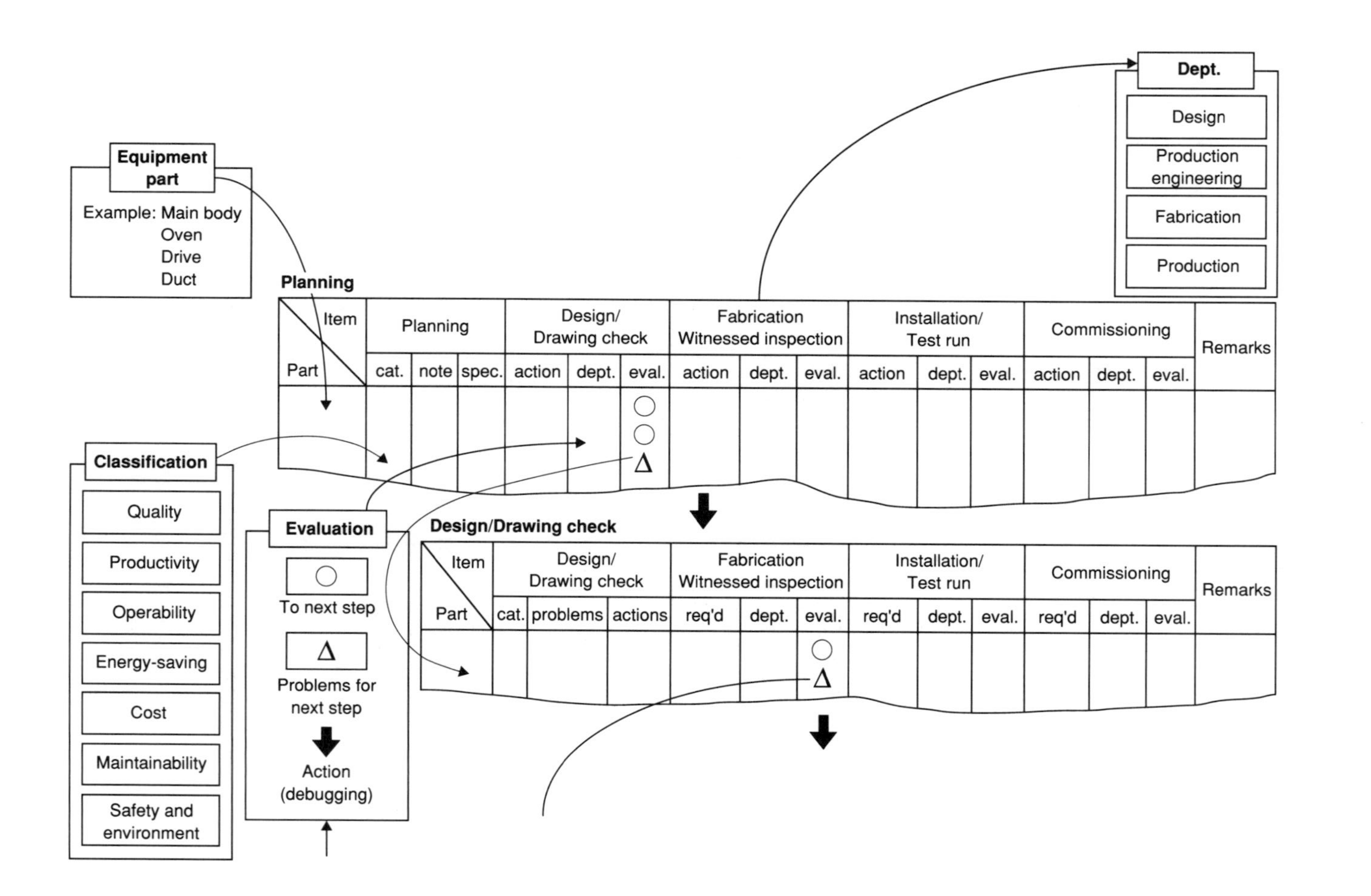
Equipment part
Example: Main body
Oven
Drive
Duct
Planning
Item
Part
Planning
cat.
note
spec.
Design/ Drawing check
action
dept.
eval.
Fabrication Witnessed inspection
Installation/ Test run
Commissioning
Remarks
Dept.
Design
Production engineering
Fabrication
Production
Classification
Quality
Productivity
Operability
Energy-saving
Cost
Maintainability
Safety and environment
Evaluation
To next step
Problems for next step
Action (debugging)
Design/Drawing check
problems
actions
req'd

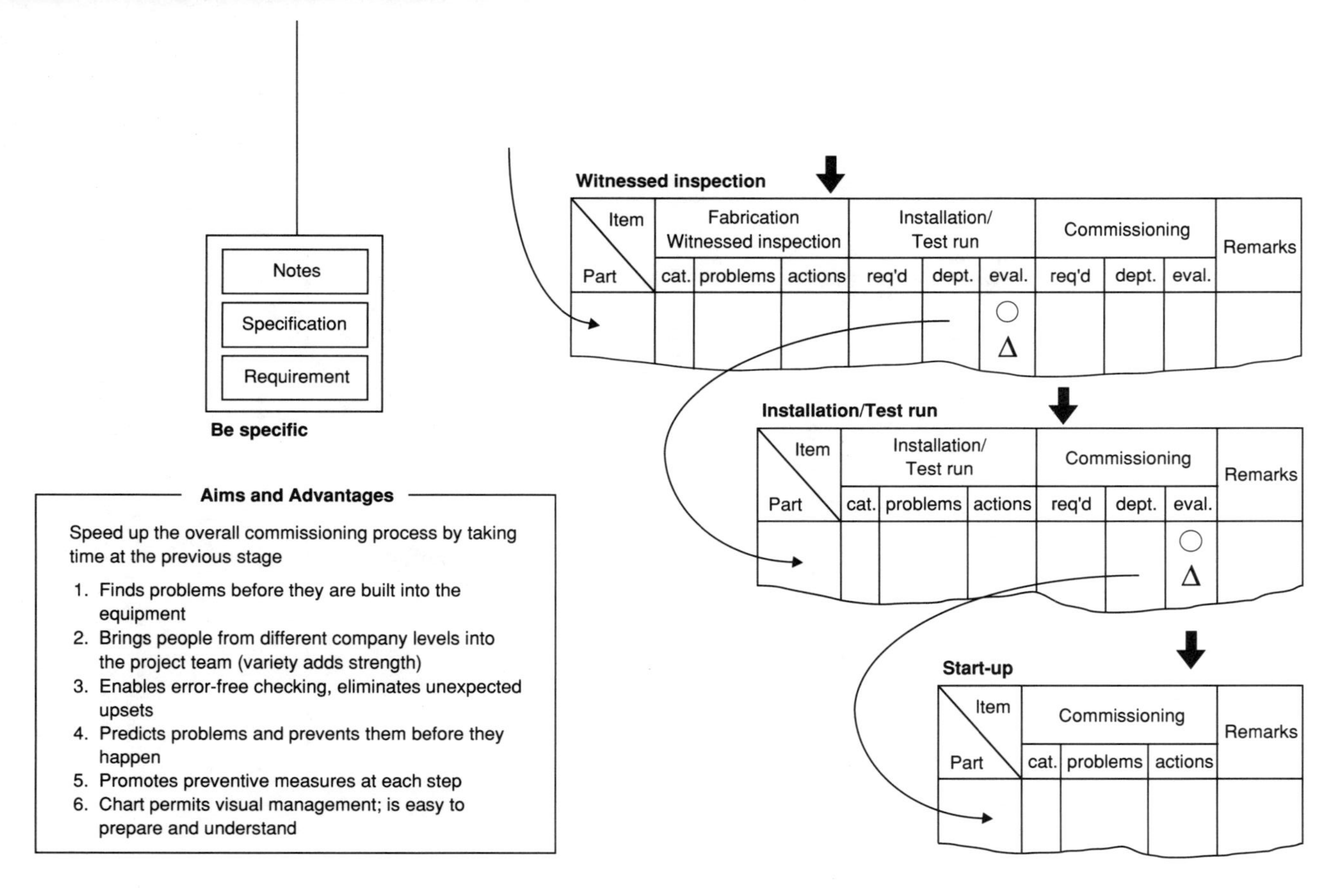

Figure 2-8. Step-by-Step Preliminary Evaluation Chart

Background

In recent years, consumer demand for a smoother ride has pressured automobile manufacturers to reduce wheel vibration (see Figure 2-10). Although each wheel consists of two main components — the rim and the disk — the shape of the rim determines the amount of vibration. Figure 2-11 is a diagram of the rim-manufacturing process. The process that most influences vibration in the finished rim is the sizing (expander) process.

In this example, the wheel manufacturer introduced a new expander machine to improve the sizing processing. The addition of a new machine in the rim manufacturing line required a startup procedure to switch from the old expander to the new one. The step-by-step preliminary evaluation chart was used to plan the introduction of the new expander.

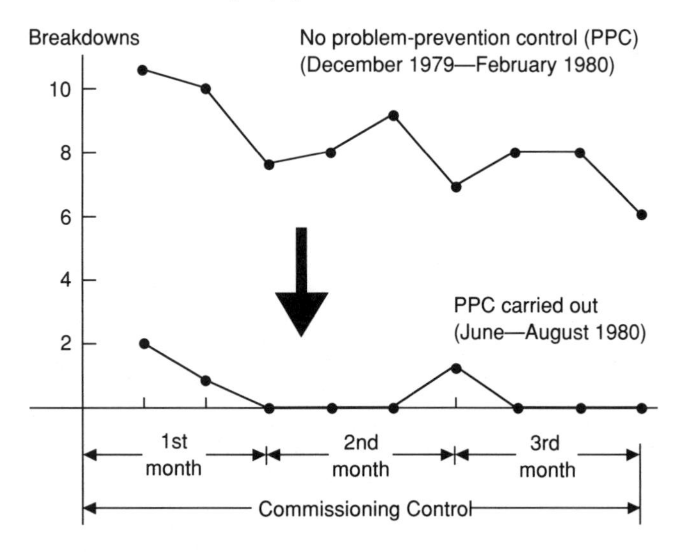

Figure 2-9. Reduction of Early Breakdowns (Comparing Similar Types of Equipment)(Tokai Rubber Industries, Ltd.)

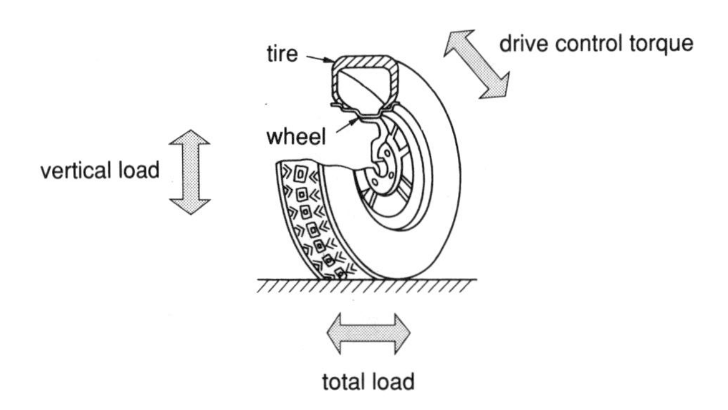

Figure 2-10. Automobile Wheel

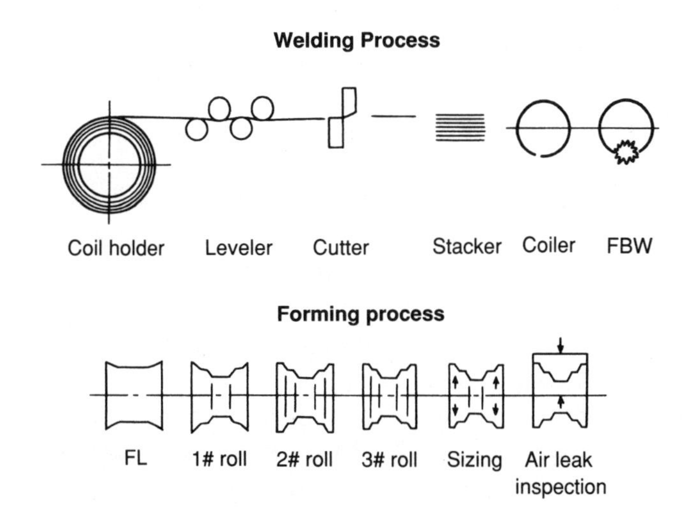

Figure 2-11. Diagram of Rim Manufacturing Process

Outline of Target Equipment

Figure 2-12 presents an outline of the expander machine's structure. As shown in the figure, the rim, which has been prepared by another forming process, is set alongside the expander's reference block, which helps ensure precise positioning,

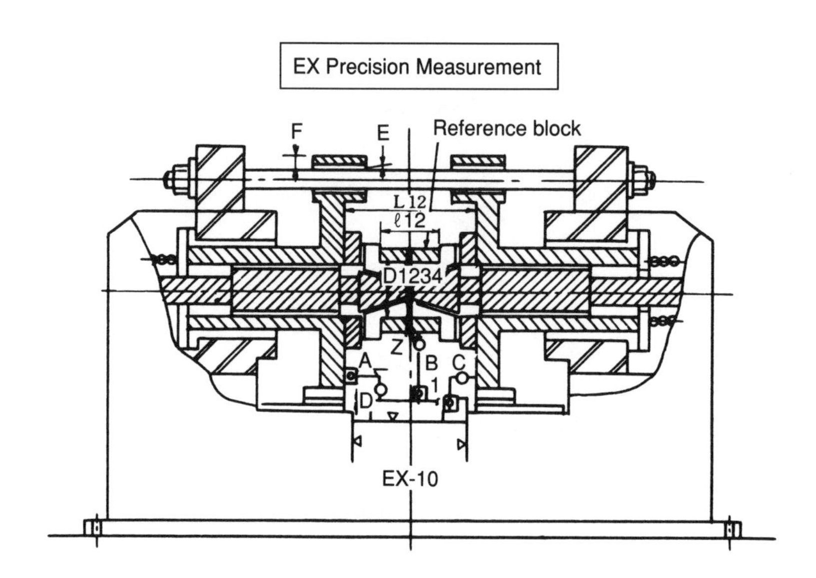

Figure 2-12. Outline of Expander Machine Structure

and then is pressed into its final form. The areas marked with various capital letters in the figure indicate measurement marks that are also used to ensure precision. The right side of the cone at the center of the expander machine moves to the left (and the left side moves to the right) to press open the segment into an expanded form.

Refer to the step-by-step preliminary evaluation chart shown in six parts in Table 2-4 to see how the project team controlled the development process from planning to installation.

COLLECTING AND USING MAINTENANCE PREVENTION (MP) DATA

The Japan Institute of Plant Maintenance defines maintenance prevention (MP) as the use of the latest maintenance data and technology when planning or building new equipment to promote greater reliability, maintainability, economy, operability, and safety, while minimizing maintenance costs and deterioration-related loss. Put simply, MP means making equipment that is designed from the start for easy maintenance and trouble-free operation.

MP Activities Help Prevent Problems during Commissioning Control

When developing new equipment, the design, fabrication, and installation stages may proceed smoothly, but at the test run and commissioning stages, numerous abnormalities may emerge to preclude smooth operation. After much effort on the part of operations and maintenance engineers, the equipment eventually operates normally, but even then, it often needs extra inspections, adjustments, lubrication, and cleaning to keep running. With all the time spent inspecting, maintaining, and repairing the new equipment, the net result is a high level of downtime and mounting frustration.

Problems originating at the design and fabrication stages have always been unwelcome during the commissioning stage, as is having to take various remedial measures to address them. This is particularly so now with rapid technological advances, larger equipment models, and faster, more highly automated equipment.

The key to preventing such problems is to ensure that all processing and operation requirements of the new equipment are fully incorporated in the equipment requirements used by the equipment design engineers. The best approach is to avoid buying equipment technology developed by another company's engineers and, instead, to develop the equipment in-house. This takes advantage of the technology accumulated over years of hard work by the company's own production, equipment, and maintenance engineers, and uses the lessons of previous internal successes and failures. This approach, more than any other activity, determines the ease or difficulty of subsequent production maintenance.

Table 2-4A. Step-by-Step Preliminary Evaluation Chart (Part 1, Planning)

Item	Planning			Design and drawing check		
	Category	**Points of interest**	**Specification**	**Requirements**	**Dept**	**Eval**
1 Pressure	Main specs	IHI top roll pressure set to maximum for modification	Pressure = 150 tons	Confirm specifications	Equipt & eng	○
2 Cycle time		5.0 seconds/unit		"	"	○
3 Path line		Match with existing machine	FL + 1,100	Check dimensions in drawing	"	○
4 Segment		Use with existing machine		Check fit dimensions	Prod eng	○
5 "	Easier die setup needed	Enable 1-piece attach/detach		Check drawing	Equipt	○
6 Adjust diameter		Electric digital display		Check drawing	"	○
7 Anchor holes, size, pitch		Match with existing machine		"	"	○
8 Hydraulic unit		Anti-vibration	Made by Tokyo Keiki	Separate motor pump from tank	"	○
9 Grease supply pipe	Maintainable	Make lubrication easier	Collective style	Check specs manual	"	○
10 Pressure cylinder	"	Consider maintainability	Attach to exterior of main unit	Check drawing	"	○
11 Process precision (product vibration control)		$\bar{x} = 0.25$ $\bar{x} = 30 = 0.4$				○

Fabrication and witnessed test			Installation and test run			Commissioning			
Action	**Dept**	**Eval**	**Requirements**	**Dept**	**Eval**	**Requirements**	**Dept**	**Eval**	**Comments**
Confirm by checking formed workpieces 5.5-JJ X 14, 60 kg/mm 2.9-ton unit or similar unit	Equipt/ prod eng		Confirm by checking formed workpieces Similar part	Equipt/ prod eng			Equipt		
Cycle-time measurement	Equipt	○	Cycle-time measurement	Equipt		Check productivity			
Check item	"	○	Adjust when fastening	"	○				
Check item	Prod eng	○							
"	Equipt	○							
Check item	"	○	Check operation	Equipt	○				
"	"	○							
"	"	○							
"	"	△							
"	"	○							
Measurement	Prod eng	△	Check: $\bar{x} = 0.25$ $\bar{x} = 30 = 0.4$	Prod eng					

Table 2-4B. Step-by-Step Preliminary Evaluation Chart (Design and Drawing Check)

Item	Design and drawing check		
	Category	Problems	Actions
1 Hydraulic pipe swivel joint	Maintainability	Risk of oil leak (leaks occur at other factories)	Use high-pressure hose for oil pipe
2 Segment attachment site	Quick setup	Need to protect faucet to support single-unit exchange method	Insert hardened rings in faucet (both sides)
3 Digital display for diameter calibration	Reliability	Wire-reel method lacks reproducibility and reliability	Consider other methods (e.g, gear coupling)
4 Lever strength	"	Cracking accident occurred at KWK (problem in material)	Check material
5 Plunger guide suppression plate	"	Fastening bolt breakage (at other factories)	Change from 6 M16 bolts to 12 M20 bolts
6 Head	"	Head wandering	Install device to stop head wandering
7 Interior inspection	Maintainability	No inspection window	Install an inspection window
8 Chute at outlet side	Easier die setup	Consider easing setup	Use air valve for opening and shutting
9 Feeder	"	"	Revolving type
10 Diameter calibration	Maintainability	Risks of unstable diameter dimension and breakdown of diameter calibrator	Interlock to prevent operation when under pressure
11 Tie rod bushing		More thorough prevention of alignment slippage	Extend at centered machine
12 Confirm machine's accuracy level		Checkpoints, numerical values, and measurement methods are unclear	Clarify items
13 Path line		Match with existing machine	FL + 1,100
14 Hydraulic unit	Maintainability	Prevent vibration-caused leak	Separate motor, pump, and tank
15 Segment		Change the material	Change to SCM445
16 Segment slide plate		Improve durability (especially abrasion resistance)	Use Hitachi Metal's FA381

Fabrication and witnessed inspection			Installation and test run		
Requirements	**Dept**	**Eval**	**Requirements**	**Dept**	**Eval**
Check item	Equipt	○			
"	Prod eng	○			
Check gear-coupling system	Equipt	○	Check operation	Equipt	○
Mill sheet confirmation	"	--			
Check size and quantity	"	○			
Check operation	"		Check operation	Prod eng/ equipt	
Check ease of inspection	Maint	○			
Check item and operation	Prod eng/ line ops	○	Check operability	Die setup	○
"	"	○	"	"	○
Check operation	Maint	○	Check operation	Maint	○
Check item	Equipt	○			
Check precision	Equipt/ prod eng	△	Check precision	Equipt/prod eng/maint	△
Check item	Equipt	○	Connect to existing chute	Equipt	○
Check item	"	○			
"	Prod eng				
"	"	○			

Table 2-4B. cont'd.

Commissioning			Comments
Requirements	**Dept**	**Eval**	
1 Check for lubricant leak	Equipt/maint	○	
2			
3 Check operation	Maint/line ops	△	
4 Periodic check	Maint		
5 Check for breakage	Line ops/ maint	○	
6 Check operation	Line ops/prod eng/ equipt		Left room for additional equipment in future (8/18)
7			
8 Evaluate ease of die replacement	Die setup	△	
9 Evaluate die setup	"	○	Fabricate new feeder based on 8/18 plan
10			
11			
12 Check precision	Equipt/prod eng/ maint	△	
13 Check flow condition	Equipt	○	
14			
15			
16 Evaluate abrasion resistance			

Table 2-4C. Step-by-Step Preliminary Evaluation Chart (Part 2, Design and Drawing Check)

Item	Design and drawing check		
	Category	**Problems**	**Actions**
1 Check interlock with existing forming roll	Produc-tivity	Insufficient terminals for interlock with existing equipment	Prepare interlock and input/output terminals
2 Operation setup circuit	"	No externally connected emergency stop terminal	Get externally connected emergency stop terminal
3 Feeder operation switch	Maintain-ability	Cannot use limit switch to prevent lubricant-related problems	Use proximity switch
4 Pressure main-tenance time	Die setup	Analog setting device means precision varies from operator to operator	Use digital setting device
5 Cone expansion/ contraction switch	Operability	Risk of operation error (pressing wrong button)	Install ON/OFF switches
6 Relay circuit box for interior wiring	Mainte-nance	Cannot install at operator side of machine	Install on hydraulic unit side
7 Reduction of oil-level problems	"	Difficult to hear alarm	Install pilot lamp on operator panel
8 Hydraulic pump start button	Operability	Does not operate on EX panel	Install on existing central operation panel
9 Operation power key switch	"	"	"
10 Ram forward position detector switches	Quality	No forward position detector switches	Install limit switches (left and right)
11 Feeder inching operation	Die setup	Inching required during die setup	Repair circuit enabling inching operation
12 Hydraulic unit's heater switch	Maintain-ability	Switch is left on all the time (year-round)	Install ON/OFF switch on operator panel
13 Timer for activa-tion of feeder movement	Die setup	Contained in sequencer, cannot be adjusted at the line	Install outside of sequencer
14 Normal stop	Operability		Set up for single-cycle stop
15 Operation circuit diagram	Operability	Do not use alphanumerics for wire numbers	Use numerics only
16 Device to be used	"		Check design standards manual

Fabrication and witnessed test			Installation and test run		
Action	**Dept**	**Eval**	**Action**	**Dept**	**Eval**
Required input/ output terminal	Main-tenance	○	Link with existing machine	Maintenance	○
Check item	"	○	Check operation		○
Check item	"	○	Check operation	"	○
"	Equip-ment				
"	"	○			
Check operation	Main-tenance	○	Check operation	Maintenance	○
Check position where item is attached	"	○			
Check item and operation	Main-tenance	○	Check operation	Maintenance	○
START button activates two machines at once	Equip-ment	○ }	Install on central operator panel and check operation	Maintenance	○
"		○ }			
Check item and operation	Equip-ment	○	Check operation	Maintenance	○
Check inching operation	"	○	"	"	○
Check item and operation	Main-tenance	○	"	"	○
"	Main-tenance	○	"	"	○
Check operation	"	○			
Check item	"	○			
"	"	○			

Table 2-4C. cont'd.

Commissioning			Comments
Action	Dept	Eval	
1 Check linked operation with existing machine	Maintenance	○	Display relay
2	"	○	OMRON MY- 4N$_A$ C100V
"	"	○	
3 Check for operation defect	Maintenance	○	OMRON A.C100V
4 Check ease of setting	Line operations	○	OMRON H$_5$ B-2D
5 Check operation	Maintenance	○	
6			
7 Check operation	Maintenance	○	
8			
9			
10 Check operation	Maintenance	○	
11 "	Line operations	○	
12 "	Maintenance		
13			
14			
15			
16			

Table 2-4D. Step-by-Step Preliminary Evaluation Chart (Part 3 Fabrication and Witnessed Inspection)

Item	Fabrication and witnessed inspection	
	Category	**Problems**
1 Grease pipe on exit side	Maintainability	Pipe position is awkward; pipe is difficult to clean
2 Air pipe on exit side	"	
3 Air connection point for die replacements	Die setup	No air connection for impact wrench used to remove and attach frame during die replacement
4 Grease pipe	Reliability	Grease pipe divided in middle, creating risk of insufficient grease supply
5 Inspection window	Maintainability	Inspection window bolted on and difficult to use
6 Counter for diameter calibration	"	Reliability
7 Limit switch to detect catcher section's load	Productivity	Risk of inadequate flow due to low lever position
8 Limit switch to detect full setback of feeder	Maintainability	Maintenance difficult because limit switches are attached on inside
9 Proximity switch for feeder	Reliability	Detector plate is too thin and may bend
10 Relay terminal box	Maintainability	Bolt-tightened type; difficult to maintain
11 Chute width adjustment	Die setup	Adjustment difficult due to lack of scale
12 Central part of main unit	"	Not covered; die replacement bolts can fall into main unit
13 Air pipe at outlet port	"	Not covered; operator can step accidentally on air pipe
14 Feeder hose	Reliability	Hindrance to hose compatibility
15 Rim catcher hose	"	Interferes with main unit
16 Tightening bolts		
17 Rim catcher	Die setup	Rim catcher gets in the way when replacing die from the rear side
18 Device nameplates	Maintainability	Electromagnetic valve and limit switches need name plates
19 Relay	Reliability	No fall-prevention devices for LY and MY types
20 Hydraulic unit terminal box		Error in wire numbers
21 Grease nipple	Maintainability	No indication of where grease is going
22 Pressure gauge	"	No indication of use range
23 Current gauge	"	"
24 Main cylinder	Safety	No safety cover for movable parts
25 Link pins		No blind cover at two places on rear side

Table 2-4D. cont'd.

Fabr. and witnessed inspection	Installation and test run		
Requirements	**Actions**	**Dept**	**Eval**
Lower part that protrudes at top	Check implementation of action	Equip/prod eng	○
Lower 3-point set (FRL) side	"	"	○
Install air outlet (3/8 collar) on workpiece insertion side	"	"	○
Install one for each grease nipple	"	"	○
Install hinged inspection window	"	"	○
Install clip lock for inspection window	"	"	○
Exchange for shorter lever	"	"	×
Keep as is to avoid protruding above space limits			
Use steel plate having 4.5 t thickness	"	"	○
Replace with hinged type (or remodel existing one)	"	"	×
Attach commercial scale	"	"	○
Attach a cover	"	"	○
Attach a cover (after installing machine)	"	"	○
Attach a hole band	"	"	○
Separate from main unit	"	"	○
Line up marks after tightening	"	"	○
Die replacement done mainly done from front side; check condition after installation	—		
Attach a plastic name plate	"	"	×
Install antifall device	"	"	○
Fix wire numbers	"	"	○
Attach name plate indicating grease supply destination	"	"	×
Use color coding to indicate use range	"	"	○
"	"	"	○
Attach push-on type of divided cover	"	"	○
Attach blind cover	"	"	○

Installation and test run			Comments
Requirements	Dept	Eval	
Exchange for proximity switch	Equipment and production engineering	○	
Replace with hinged type	Equipment and production engineering	○	Finished on 2/17
Attach a name plate	Manufactured by Watanabe Steel	○	Finished on 3/14
Attach name plate	Manufactured by Watanabe Steel	○	Finished on 3/14

Table 2-4E. Step-by-Step Preliminary Evaluation Chart

Item	Installation	
	Category	**Problems**
1 Anchor bolt	Installation	Anchor bolts from current model are being used in new model with different base plate thickness and different fastening methods. As a result: a. Bolt is too long and obstructed by main unit's bolt openings. (Nut-fastened bolts in the current model are not obstructed) b. Bolts are only partially screw-threaded, so a collar must be added if nuts are to be used c. Screw-threaded part of bolt is too short to enable use of double nuts
2 Feeder section	Productivity	Feeder can't be formed into same shape as chute, therefore: a. Limit switch bar that checks load on rim catcher acts as a brake and stops workpiece b. Unless top of feeder attachment is cut off, it will not fit angle of chute, and workpieces cannot pass through
3 Rim catcher	Die setup	Dimensional variance in rim catcher's fastening section depending upon the rim catcher's size. Cannot be tightened with key
4 Precision gauge		Measurement is obstructed by grease pipe that runs parallel to front of ram and dial gauge
5 Main unit's central cover		Cover blocks feeder when feeder is raised

Installation	Commissioning			Comments
Actions	**Requirements**	**Dept**	**Eval**	
1. Cut bolt length to about 100 mm 2. Fabricate collars (two vertical collars for each bolt) 3. Spot weld bolt after attaching nut				Finished during installation " "
1. Replace limit switches with proximity switches 2. Cut about 20 mm from the top of the feeder attachment	Check flow conditions	Equipment engineering	○	Finished during installation
				"
	Modify slide stopper for measuring	Made by Watanabe Steel	○	Finished on 2/17
Obstructing areas cut off with gas				Finished during installation

Table 2-4F. Step-by-Step Preliminary Evaluation Chart

Category	Problems	Actions	Comments
1 Feeder	Feeder can be raised only part way. No electrical or hydraulic problems	Where cylinder is disconnected, there are not enough shims for the head-fastening section on the lot side. So bolt is fastened too tightly and cannot be moved. Use super-short packing to adjust total shim width	Correct shim adjustment done by Tokyo Keiki on 1/13
2 Rim catcher	When replacing with 4T × 16 rims, height control locks vertically and cannot be adjusted properly	Raise rim catcher's vertical adjustment mechanism up 40 mm	Obstruction between warm start linkage shaft and rim catcher bracket (design error)
3 Counter	a. Damaged lots of counters and connectors b. Upper and lower limit switches do not operate; will not move in direction of rim lead's inner diameter	Modify interior electrical connections to keep limit switches from operating immediately	Replaced end of March when replacement parts arrived
4 Feeder	Large shock to the return side	Tested return-side deceleration valve operation but gave up because timer was needed for fine-tuning	Plan to exchange with return-side valve for one with head cushion during vacation week in May
5 Ram cylinder trunnion	Grease does not reach inside and leaks from the outside	There are no grease holes in the metal; need to make grease holes	Finished on 3/4

Roles of Equipment Design (and Fabrication) Engineers and Maintenance Engineers

The following three factors determine the quality of MP activities:

- design engineers' skill and sensitivity
- quality and quantity of technical data
- usefulness of technical data

Design engineers do not refine their skills through other people. They improve their skills through practice — by exercising their own creativity and initiative in their work. Nevertheless, engineers cannot rely completely either on theories developed behind a desk or on their own experiences. Consequently, all in-house engineers who do equipment-related work should pool their talents in gathering technical data to bring more reliability and maintainability into the equipment design. Once such data have been gathered and organized, the group should use the data to establish a set of technical design standards.

Often, the technical improvement data gathered by maintenance engineers during daily preventive maintenance activities — data that can improve the equipment's reliability and maintainability — are never put to use. Maintenance engineers sometimes fail to organize and present to design engineers the technical data pertaining to reliability or maintainability that should be considered at the equipment design and fabrication stages. Sometime the equipment designers themselves fail to gather, organize, and use similar technical data.

The maintenance engineers must not only supply maintenance data as feedback to the design engineers, they should also actively support the design engineers. The design engineers, in turn, should assume greater responsibility for the equipment they design — even after it leaves their hands — by keeping tabs on the life of the equipment. They will then be better prepared to develop technical solutions to the various bugs that appear in the equipment after it is running. Continuity in the mutually supportive relations between equipment design and production engineers can thus be maintained. Together, the engineers can create technical standards for equipment design that result in more appealing, reliable, and maintainable equipment models. Such technical cooperation is an essential ingredient for effective MP activities.

MP Data Collection System

Figure 2-13 shows a system for MP-related feedback and standardization. The feedback encompasses all types of relevant data concerning safety, quality, maintenance, production engineering, and other areas. These data were gradually accumulated and stored in a log file. Equipment data on a wide range of equipment were developed into machine tool design standards, safety standards as part of the companywide feedback, and standardization activities.

Problems Related to MP Data

Figure 2-14 is a simple flowchart illustrating how MP data are circulated around the company as feedback. As shown in the figure, there are five areas where problems typically arise as MP data are sent to various departments.

1. Maintenance log. People in the maintenance department are usually required to fill out a maintenance log sheet for every breakdown repair, preventive maintenance measure, or other maintenance operation. Equipment designers should analyze these records to develop preventive actions and incorporate those actions in the next generation of equipment. Problems arise when the contents of the maintenance log are inadequate.

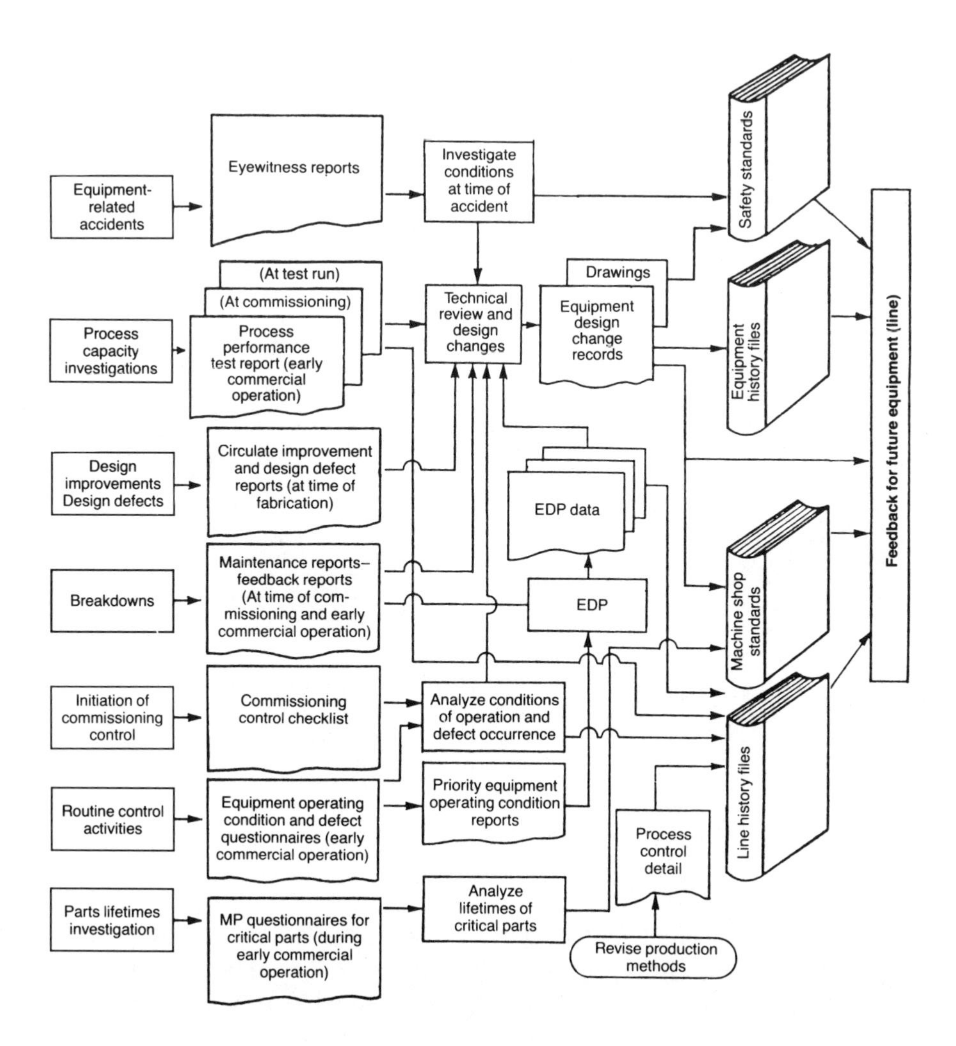

Figure 2-13. Feedback and Standardization of MP Data (Nippon Denso)

For example, the log might describe the repair but not the cause of or reason for doing the repair, or how effective the repair was in eliminating the problem.

2. ***MP feedback data.*** Even when MP data are passed on from the maintenance department to the equipment design department, the contents are not always sufficiently coherent, or

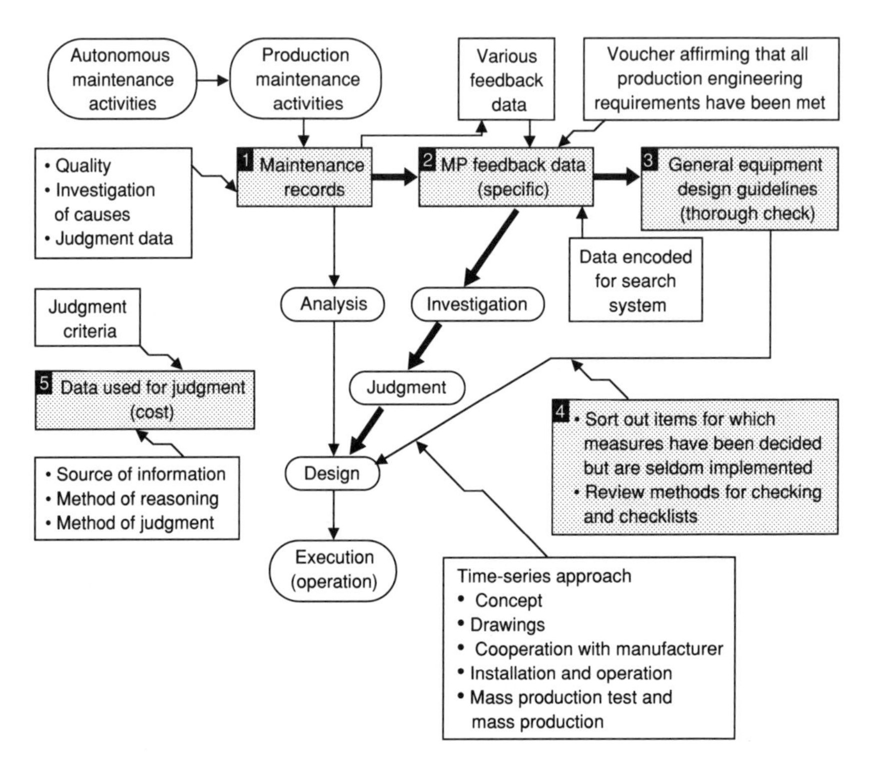

Figure 2-14. Five Problems Related to MP Data

the cost and effect factors may be too vague. The equipment designers may thus find the data impossible to use. (For additional discussion of this subject, see pages 93-95). Unless the feedback data are organized into coded categories, for example, it becomes difficult to search the data, and it may become totally useless.

3. ***Design standards and guidelines.*** Ideally, all MP feedback data should be analyzed for common problems and solutions and then made generally applicable as equipment standards or guidelines. Quite often, however, the maintenance data are put to use only once as MP feedback data. Even

when such standards or guidelines are written, they are often stated so vaguely that the equipment designers must interpret them according to their own criteria to make them useful.

4. Checklists. Difficult as it is to have MP feedback data reflected in equipment design standards, it is even more difficult to make sure those new standards and guidelines are reflected in the next set of equipment designs. Checklists are useful generally as a means of finding areas where standards are missing from the design, but they are far from foolproof. Whatever omissions eventually become evident should be incorporated to review the various checking methods and checklist contents. Start with the design-concept stage and proceed chronologically to the mass-production stage to make sure that the checklist at each stage covers all pertinent items.

5. Judgment of usefulness. When a vague method is used for judging how well the various feedback data are reflected in the equipment designs, the vagueness leaves room for different equipment designers to judge the design differently. To eliminate such ambiguity, establish rules for stating the source of the information, the method of reasoning, and the method of judging usefulness. You should also organize cost data and other relevant data.

Maintenance Data Feedback

Factory operations and maintenance departments often carry out improvements on existing equipment to prevent breakdowns or quality problems, or to boost advantages such as operability or maintainability. Naturally, all such improvement measures should be reported as feedback to the equipment design department so that the improvements can be incorporated in the next generation of equipment. As mentioned earlier, however, various problems in the flow of feedback can obstruct this goal.

With these problems, even the most reliable flow of feedback from the operations and maintenance departments does not necessarily lead to equipment design improvements. Reasons for this failure can be found both in the operations and maintenance departments and in the equipment design department. Except when designers simply neglect to make use of the feedback data, their main problem is the inability to select, out of a large quantity of data, those they need for designing improvements. This problem has two chief causes:

- The MP measure described in the feedback data is not expressed in a way that the equipment designers can readily understand.
- Even if the MP data are coherent to the equipment designers and seem effective, the corresponding design improvement may incur too many costs, or the designers may lack data indicating how effective the improvement will be.

The first cause is a lack of technical training on the part of operations and maintenance personnel. The second cause is the equipment designer's failure to make their needs known to those same departments. To resolve such problems, equipment designers should spell out their needs, and they should also describe the proper format for feedback data.

Figure 2-15 is an example of a completed MP proposal form. The abnormalities, actions, and expected effects are all quantified as much as possible to make the data easier for equipment designers to use. When the equipment designers provide the operations and maintenance departments with a form such as this to guide their MP data feedback activities, the improvements they make on existing equipment are more cost-effective because the departments are spared the labor- and loss-related costs of creating their own MP proposal forms to fill out. The result is a higher effect-to-cost ratio and more technically advanced MP feedback activities. The highly quantified proposals that such established forms help produce have a much better chance of being used by equipment designers to improve future designs.

Standardization and Use of MP Data

When new equipment designs or existing equipment modifications fail to achieve the expected levels of reliability and maintainability, the chief reason is usually the inadequate accumulation and dissemination of related technology.

Standard Data

If you simply gather and file away maintenance reports without further processing, they are practically worthless. Such data gathering cannot be valued as an accumulation of technology. Conversely, technological know-how that exists only in the minds of veteran engineers and is not taught to newer, less experienced engineers is also of little value. Accordingly, a guidebook that presents a standardized compendium of maintenance and other experience-based data will help equipment designers improve their equipment development skills and avoid common pitfalls.

Standard Parts

One of the greatest obstacles to efficient maintenance is equipment with a wide variety of parts. When parts that perform the exact same function come from different manufacturers or in slightly different models, they can result in larger spare parts inventories and the potential for long delays due to parts shortages. Repair related errors are also easier to make.

To solve this problem, design standards should keep the variety of parts to a minimum (by combining functions in fewer parts, for example) as long as it does not sacrifice any significant

Factory: 2F
Job: Painting
Process: Undercoating

Equip. cat.: Lifter 0024
Machine no.: HCV495
Name: #3 drop lift

Equip. rank: (circle one)
A = primary heavy resins
B C D E

Application in other equipment? No / Yes Example: ____

Abnormality/Defect:

3D/L motor burnout process

Date	Down	Abn. units	Date	Down	Abn. units
9/23/83	34"	J: 3 units Polish: 4 units	12/19/86	13"	Polish: 3 units
11/17/83	17"	Polish: 5 units	9/24/87	41"	J: 3 units Polish: 8 units
3/30/85	107"	J: 19 units Polish: 6 units	1/16/88	17"	Polish: 4 units

Results (P-M analysis):

Insulation deteriorated due to coil heat from high-speed side of D/L lifter motor

	Capacity	Rating	Measurement
High sp	3.7 kw	7.4 A	65.0 A
Low sp	0.62 kw	8.4 A	7.5 A

As shown above, inadequate capacity results in intermittent operation with a steady current about nine times the rated steady current. This has resulted in repeated burnout of the motor

Goals (check box)

- ☐ Intermittent operation
- ☐ Equipment breakdown
- ☐ Quality improvement
- ☐ Simplified cause search
- ☐ Simplified maint. tasks
- ☐ Operability
- ☐ Ease of handling
- ☐ Safety
- ☐ Economy
- ☐ Other

Effects of breakdowns	86	87	88	**Goal**
Down time	13	41	17	0 min
Labor hours	217	683	283	0 hrs
Energy				
This unit's downtime	13	41	17	0 min
Severity	0.1	0.2	0.1	0
Breakdowns	1	1	1	0
Quality defects	3	11	4	

Ease of repair				**Goal**
Labor hours to find problem	10	35	15	0 hrs
Labor hours to fix problem	12	12	12	10
Outside expenses*	$600	$600	$600	0

Action to prevent recurrence

1. Raise motor's capacity (higher costs)
2. Suppress heat by using low speed only (lower costs)

⇩

Burnout prevention via improved maintenance

1. 3-second increase in 3O/L driver roller's cycle time
2. Use #2 roller flight accelerator plate to even out the feed pitch ... 6 seconds

Make above improvements and use low speed only

Ease of maintenance				**Goal**
PM frequency	2/	2/	2/	0
PM labor hrs	30m/	30m/	30m/	0
Outside expenses*				

***Outside expenses**

Part/model	**No.**	**Type**	**Category (1-2)**

Results

(Results of investigation by departments submitting proposals and carrying out planning)

- Insufficient consideration given to equipment use conditions in relation to motor's characteristics (for example, inadequate investigation for moving or modifying equipment)
- "Items to be studied" for next round of equipment design need to be incorporated into standards manuals

☐ Use
☐ Use with other method
☐ Don't use

Standardization

Name: conveyor design standards
Standard No.: TPD-UC-1910
Planning dept.: No. 2 prod. engineering, coating technology section
Date: July 10, 1987
Signed: Nakamura

Figure 2-15. MP Proposal Form

technical advantages. A wide variety of parts is not always disadvantageous, since it attests to broad-range study by the design engineers who selected the parts. Be wary, however, of too much variety in equipment parts or too much emphasis on novelty.

Table 2-5 shows a design standard for retaining screws. In this example, the company did not simply adopt the national standard, but instead, customized it to address the problem that originally prompted creation of the standard. The standard's application range and relevant design and use methods are described in simple and clear language.

Design Standards and Checklists for Debugging

Standards that are too long and wordy are more difficult to search and understand, which makes them harder to use. Standards are not carved in stone; once they are written, you should review them regularly and update them with new information so they are always fully applicable as design guidelines. Effective standards will help prevent problems when you use them as the basis for debugging checklists. At every stage from design to fabrication, installation, test run, and commissioning, you should use the most appropriate standards as items in a stage-specific checklist. Such debugging checklists can help uncover design and fabrication errors and other abnormalities earlier, when it is easier to fix the bugs. The commissioning stage is already too late to concentrate on debugging. Repair at that stage incurs much higher costs and makes it difficult to achieve the primary objective of commissioning control, which is to get the equipment operating normally as soon as possible.

1. Example of checklist used at design stage. Table 2-6 shows a checklist used at the design stage. In addition to general items that relate to design tasks, this company also uses a checklist for preventing design error. One of the items on the main checklist asks whether the design error prevention checklist is also being used.

Table 2-5. Design Standard for Set Screws

Machine Standard: K-3702
Retaining Screws

1. Reason for preparing this standard:

   A recent series of equipment breakdowns was caused by fixed subassemblies shifting as a result of loosened set screws. This standard is intended to prevent this problem.

   Example:

   CM-5 Shifting of sprocket of elevator unit led to abrasion and breakdown (July 23,1974)

   CM-D Shifting of TR part led to abrasion and breakdown (October 5, 1974)

2. **Scope of application:**

   This standard applies to set screws used with the parts described below:

   1) Parts requiring no further adjustment after initial positioning
   2) Permanently fixed parts that might cause a mechanical breakdown if they shifted

3. **Design and method of use:**

   The present standard specifies set screws of the hex socket head type following JIS.

   1) When removal is unnecessary... hex socket head set screws with knurled cup point
   2) When removal and adjustment are required... hex socket head set screws with cup point or cone point

Note: Since hexagonal socket head set screws are suitable for the scope of applications listed in para. 2, they have been limited to these. For parts outside this scope, other types of screws indicated in JIS may be used. See reference materials.

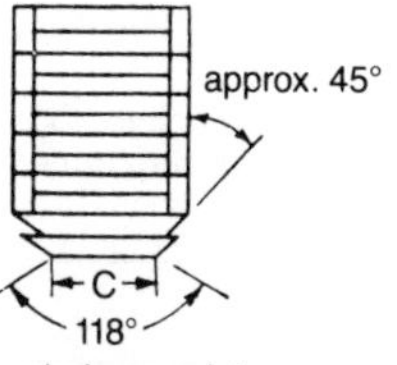

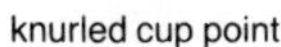
knurled cup point

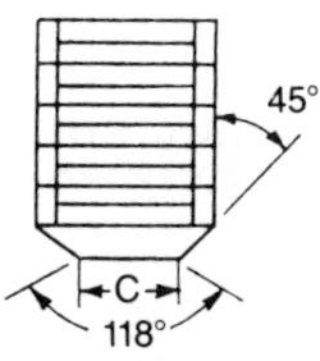

cup point

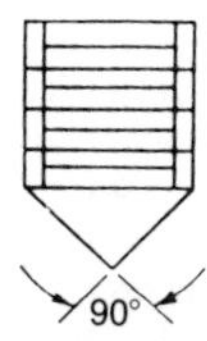

cone point

2. ***Example of checklist used at acceptance inspection stage.*** Table 2-7 shows a set of standards for a lubricating device. In this example (which is limited to lubricating devices), the checklist asks whether standard parts (based on the design standards) are being used. It also lists the most important items, such as installation, pump assembly, and pipes, to be checked during the acceptance inspection of the completed equipment.

Table 2-6. Design Checklist (Fuji Photo Film)

Design standard: Y-0302
Design Checklist (Preconditions) 1. Tackle the task positively. Use sound technical knowledge and a scientific approach to achieve the best results. 2. Work in active collaboration and cooperation with other departments concerned.
Check Details
I. Planning and design: 1. Do you know the object of the design? (reason, conditions set by originator of design request relevant conditions) 2. Are the design procedures appropriate? (method of execution, completion date, priority schedule planning) 3. Is there satisfactory contact with the originator of the design request? 4. Has the site been thoroughly investigated? 5. Are reference materials adequate? (technical data, introduction of new technology use of existing technology) 6. Are the most suitable and optimal methods and systems being applied? Has complacency been avoided? 7. Are design calculations error-free? (strength, functions, capacity) Have all problems been fully considered? 8. Are maintenance prevention considerations adequate? (Was the maintenance department consulted and did they confirm in advance; will they check the design afterward?) 9. Will the design be cost-efficient? (within budget, operating costs) Is operability good; has safety been considered? 10. Has the optical sensitivity of materials been checked? (Has a request for the photographic characteristics test been issued?) 11. Have related departments been contacted?(maintenance engineering, electrical and instrumental, packing engineering, fabrication departments, safety department)
II. Drawings 1. Have the drawings been reviewed? Are they error-free? (dimensions, number of parts, accuracy, materials, procurement of spare parts, use of checklists to prevent design errors) 2. Has microfilming been considered? 3. Have cost reduction checklists been used? 4. Have the drawings been checked and approved?
III. Purchasing 1. Are specifications of equipment to be purchased satisfactory? (use of standard documents, selection of equipment) 2. Are purchasing arrangements satisfactory? (no mistakes in the arrangements, delivery times, prices, selection of manufacturers) 3. Have vendor's estimates been thoroughly reviewed? (prices, delivery times, details)

Table 2-7. Lubrication Device Standard (Nippon Denso)

Nippon Denso Mechanical Engineering Standard DMS 1-025001 B	
Lubrication Device Standard	**Issued:** August 25, 1969 **4th revision:** March 19, 1977 **__th confirmation:** ________

4.1 Items to be performed by manufacturer

(3) In principle, the parts used in lubricating devices must be those specified in DMS standards.

4.2 Installation

4.2.1 (1) (a) Lubricating devices must be installed or protected so that they cannot be damaged by falling objects, careless material handling, or careless actions of workers.

(b) Lubricating devices must not be installed where they may overheat.

(c) Lubricating devices must be installed where they can be easily adjusted, repaired or replaced.

(d) All lubricating devices must be installed so that they do not interfere with the adjustment or maintenance of plant equipment. They must also not be installed where they may hinder normal work.

(2) Lubricating devices must not be installed in locations where operators will have to reach over rotating main shafts or tools in operation to supply oil or otherwise attend to the devices.

4.2.2 Unless specifically required by their dimensions or function, control devices must be installed between 30 cm and 180 cm from the work floor.

5.1 Installation of pumps

(1) Pumps and associated equipment must be installed in easily-accessible positions for maintenance.

(2) Pumps should be installed on the outside of lubricating reservoirs.

6.3 Piping

6.3.1 Piping joints must be designed and installed for rapid assembly and disassembly using hand tools.

6.3.2 Piping from the end of one lubrication part to the next must not be jointed on the way by welding or any other method. Joints must not be used except when required for length adjustment or assembly. Piping must also be removable without removing any plant equipment parts.

6.3.3 Piping must not be installed where it will interfere with normal operation, adjustment and repair of equipment or with replacement of lubricating devices and cleaning of oil reservoirs.

7.1 Construction of oil reservoirs

7.1.1 Oil reservoirs must be constructed so as to prevent the ingress of water or other foreign particles and to prevent oil leaks and bleeding.

7.1.2 Oil reservoirs must be constructed for easy cleaning and draining.

7.1.3 Oil reservoirs must have an oil supply port. Oil supply ports must be fitted with a strainer and have a suitable cap or cover. Methods must be devised to prevent the cap or cover from being lost.

7.1.4 Oil reservoirs must be fitted with a level gauge positioned so that the oil level can be seen from where the oil is supplied.

8.1 Filters and strainers

8.1.3 (1) Filters and strainers must be fitted with filter media which are easily replaced without stopping equipment.

Checklist examples for installation and test run stages. Table 2-8 is a design capacity checklist, Table 2-9 is a data sheet used during the test run stage, and Table 2-10 is a maintainability checklist. Table 2-8 is used to check the static and dynamic precision of installed equipment, while Table 2-9 is used to record dynamic precision measurements of the rollers. Table 2-10 is used to keep a record of detailed check items related to reliability and maintainability. This list is comprehensive; care has been taken to avoid overlooking any minor defects.

Predicting Problems and Setting up Checklists for Later Stages

Using standardized checklists such as those in Tables 2-6 through 2-10 is not enough to ensure thorough debugging. You must study the checklist contents, as well as the results of the actions taken at each stage, to predict problems and sort out priority items to be checked at subsequent stages. This will help address items that are unique to individual machines and are not covered by standard checklists.

Establishing Commissioning Control

Commissioning control comes after the installation and test-run stages. Products are manufactured while any remaining bugs are addressed with a view toward establishing stable, normal operation as soon as possible. (The installation and test-run stages are sometimes referred to jointly as the test-run stages and are considered part of the commissioning process.)

A series of checks — part of the comprehensive effort to prevent problems from being carried over into the commissioning control stage — should have been carried out already. This stage is the last chance to detect and fix errors that have escaped prediction or detection at previous stages. Frequent breakdowns

Table 2-8. Designed Function Checksheet (Fuji Photo Film)

Equipment Standard K-4412

1. **Major dimension check** (compare with dimensions specified on drawings)
   - Radius of rotation
   - Arm
   - Chucking nose
   - Dimensions of bed and position of anchor bolt holes, etc.
2. **Finished precision check** (as per precision specified in relevant drawings)
   - Horizontality of chucking nose (within 5/100)

Arm	Chucking		Unchucking	
	Unloaded	Loaded	Unloaded	Loaded

- Horizontality of chucking nose (within)

Arm	Chucking		Unchucking	
	Unloaded	Loaded	Unloaded	Loaded

- Horizontality of auxiliary rollers (within 5/100)
- Parallelism
- Horizontality — ( )
- Parallelism — ( )

3. **Assembly condition, check**
4. **Actuation check**

at this stage clearly mean that efforts at the previous stage were inadequate.

Debugging targets at the commissioning control stage should include mainly shop-floor problems related to production quality, supplying materials, and conveyance. Handing

Table 2-9. Sample Test Run Data Sheet (Fuji Photo Film)

Roller Speed Check Sheet **1) Date of measurement: November 20, 1975**

Counter speed / Roller Dia.		39.6 – 40.1 m/min		70.0 – 70.1 m/min.		(m/min.)	
		Measuring instrument A (m/min)	Measuring instrument B (rpm)	Measuring instrument A (m/min)	Measuring instrument B (rpm)	Measuring instrument A (m/min)	Measuring instrument B (rpm)
1	120	(38.6–38.7)	105–106	(67.5)	185–186		
2	318	39.3	39–40	69.2	69–70		
3	120	—	105–106	—	185–186		
4	120	39.3–39.4	105–106	68.7–68.8	185–186		
5	—	—	—	—	—		
6	80	39.3–39.4	158–159	69.1–69.2	278–279		
7	120	38.9–39.0	105–106	68.5–68.6	185–186		
8	120	38.8–38.9	105–106	69.0–69.1	185–186		
9	120	39.3–39.4	105–106	68.5–68.6	185–186		
10	100	[39.5–39.6]	126–127	69.0–69.1	222–223		
11	120	39.3–39.6	105–106	[69.3–69.4]	185–186		
12	120	(38.6–38.7)	105–106	68.5–68.6	185–186		
13	120	39.0–39.1	105–106	68.7–68.9	185–186		
14	100	39.3–39.4	126–127	69.2–69.3	222–223		
15	120	39.2–39.3	105–106	69.2–69.3	185–186		
16							
17							

( ): Minimum value
[ ]: Maximum value

control of equipment over to the operations and maintenance departments can be facilitated by training the relevant people in those departments and by providing them with maintenance standards for equipment operation, retooling, lubrication, and routine inspection and maintenance tasks.

Example of a Commissioning Control System

Figure 2-16 shows a flowchart of Yokohama Rubber's commissioning control system. The most prominent feature of this

Table 2-10. Maintainability Checklist (Fuji Photo Film)

Item	Points of interest and method of checking
1. Are screws fitted with locking aid?	Are locknuts, spring lock washers, and locking compounds in use?
2. Are welds of satisfactory strength?	Is weld overlay adequate? Examine cut welded portion with special care.
3. Are shock-absorbing devices effective?	Is there any shock that will affect parts lifetimes? Do shock absorbers work, and are they controlled?
4. Are parts adequately finished?	Is there any chance of scuffing or defective movement through inadequate finishing? Compare with drawings and modify if necessary.
5. Can parts be replaced?	Give priority to examining areas where deteriorated parts or consumable items must be replaced.
6. Are there any easily-fatigued or damaged parts?	Have any such parts or dangerous parts become obvious during test run?
7. Are any geared belts subject to pitching?	Check during test running.
8. Are positioning methods adequate?	Can positioning be secured accurately through the use of positioning notches, guides, and so on?
9. Are any parts rusting?	Check rusted parts or parts that seem likely to rust. Is surface treatment adequate?
10. Are springs properly assembled?	Are any springs subject to unreasonable strain due to assembly method, compression, or tension?
11. Are arms, brackets, and studs properly attached?	Is any bending or twisting observed during test run? Are these parts securely assembled?
12. Hydraulic cylinder assembly, oil leaks	Is there any oil leakage from hydraulic cylinder? Is assembly method as designated and of adequate strength?
13. Installation and locking of speed controllers	Are speed controllers installed properly? Are speed gauges and locks attached?
14. Roller and bearing replacement	Can rollers and bearings be replaced?
15. Are there any places where tools cannot be used?	Can tools be used in places where adjustments are required? (guides, arm positions, shearing machines, packing machines, etc.) Or are special tools needed?
16. Are covers easy to handle?	Are the safety covers of drive mechanisms and edged parts securely fixed, safe, and easily handled?
17. Is wiring securely fixed?	Is all wiring inside machinery securely fixed, out of contact with moving parts, and properly sheathed?
18. Are all cable connectors properly prevented from loosening?	Are all cable connectors firmly inserted and not loose?
19. Are brush and commutator properly contacted?	Are commutator surfaces, brush contacts and attachment methods satisfactory? Is there any slackness?
20. Are foreign particles being thrown up by gears and belts?	Is any powder or other foreign matter being thrown onto the workpiece from plastic gears, synchronous belts, etc.?
21. Limit switches	Are limit switches installed in easily visible positions? Is there proper contact with the toggles?
22. Are shafts and couplings easily replaced?	Can these be dismantled and assembled without affecting other parts or their accuracy?
23. Replacement of clutches and brakes	As above. Is the wiring securely fixed?

system is the clear definition of roles for production, maintenance, and planning (design) personnel, who act within a cooperative framework. Cooperation is the key, because commissioning control is a joint activity among these three departments.

A second major feature of the system is the clear definition of initiation and cancellation procedures for commissioning control. When you initiate commissioning control, you should already have specified your cancellation criteria, such as production capacity, stoppage frequency and effects of stoppage (downtime, etc.), and defect rate. Table 2-11 shows an example of an initiation/cancellation notice for commissioning control.

A third feature is the use of the mean time between failure (MTBF) analysis chart to record problems and improvements that take place during the commissioning control stage. This method facilitates accurate recording and control of commissioning-related data.

Establishing these features and devices does not always guarantee smooth operation. You must try to predict potential problems and manage priorities.

Example of Documentation Supporting Equipment Handover

When cancelling commissioning control and handing equipment over to the operations and maintenance departments, be prepared to provide all the necessary design documentation, including standard manuals and other materials. Table 2-12 shows a standard equipment handover sheet.

PRODUCTION SETUP PROCEDURE AND DAILY MANAGEMENT

When conducting extensive equipment modifications that result in virtually new equipment, incorporate the early

Table 2-11. Commissioning Control Initiation and Cancellation Notice

Initiation/Cancellation Notice for Commissioning Control of Mechanical Equipment

This is to give notice that commissioning control has been (initiated/cancelled) with respect to the equipment shown below.

	Initiate	Cancel
Plant manager		
Manager		
Foreman		
Person resp.		

Equipment name: automatic lathe for external finishing of pump plungers

Equipt No.: CG280

Location: injection pump shop, No. 2 production section, No. 2 machine unit, plunger finishing line;

Production: plunger production group, No. 2 machine unit, No. 2 production section

Maintenance: maintenance group, jigs and tools unit, pump engineering section;

Engineering: tooling and equipment group, pump engineering section

Details	**Initiation**	**Cancellation**
	March 4, 1980	June 5, 1980
Reason	1. New, high-cost equipment Purchase price = $40,000 2. Automatic lathe. Many problems experienced with existing No. 1 machine.	Commissioning control cancelled on achieving stable operation. 1. Work drive defective — work drive mechanism improved — OK 2. Automatic conveyor operation defective —feed screw backlash eliminated — OK 3. Output advance action defective — addition of interlock with fixed-distance sliding — OK
Measures	**Targets**	**Results**
Production capacity	24,300 units/200h	24,300 units/200h
(a) Net processing time per unit	20 sec	20 sec
(b) Operating rate	75%	76%
(c) Loading rate	90%	168%
Stoppage frequency (per unit operation hrs)	0.2%	0.2%
Stoppage downtime (per unit operation hrs)	0.3%	0.05%
Defect rate	1.0%	0.06%

Route: Engineering section keeps original; copies to all related departments

(Takahashi, Giichi: *Productive Maintenance Promotion Manual* (Tokyo: Japan Institute of Plant Maintenance, 1975)

Table 2-12. Equipment Handover Standard (Fuji Photo Film)

Equipment Handover Standard

1. Scope of application: This standard covers instrumental technical data requirements at the equipment handover to maintenance department for the purpose of smooth execution of maintenance work.
2. Data needed: As shown in table below

Technical data requirements for transfer of authority

Subject \ Data	Equipment list	Parts list	Circuit diagram	Wiring diagram	Time chart, flow chart	Flow sheet	Actuation description	Operating manual	Acceptance and test run data	External dimensions drawing, layout drawing
Single circuit	◎	◎	◎	◎			○		○	○
Control loop	◎	◎	◎	◎		○	○	◎	◎	◎
Simple sequence	◎	◎	◎	◎	○	○	○	○	◎	◎
Complex sequence	◎	◎	◎	◎	◎	◎	◎	◎	◎	◎

◎: Data that must be handed over
○: Data that should be handed over in principle but which may be omitted by agreement with maintenance department

Supplements (Additional explanation of data table)

- Subject (classification of equipment)
  1. Single circuit: simple circuit as combination of switching devices (e.g., switching , solenoid valves, etc.)
  2. Control loop: industrial instruments, position controller, etc.
  3. Simple sequences: control boxes designed in-house
  4. Complex sequences: sequences designed by outside manufacturers or their equivalent
- Data name (classification of data)
  1. Equipment list: similar to equipment log, including brief details of instrumental equipment (e.g., name, type, quantities, etc.)
  2. Parts list: parts list (all parts recorded)

equipment management policies just described into the overall setup procedure for production. The goal of these policies is to enable vertical startup, bypassing the early failure period to achieve the product quality and cost targets required of the equipment being designed. In so doing, you should also take the new product's deadline (off-line delivery) seriously. Implement all policies within the time period defined by this delivery deadline. Decide what steps will prepare the equipment for production; then, at each step, set standards that answer the following questions.

- What is being decided?
- What data are needed to make the decision?
- What tasks must be performed?
- Who can perform what tasks?
- What is the output format?

Next, calculate backward from each product's delivery deadline to work out a schedule for the completion of each product model. This gives you a production management outline to follow. Take this approach to review the following examples of preproduction setup procedures and daily management schedules.

Table 2-13 shows an outline of production setup procedures. The examples given in the table have the following characteristics:

New product development stage. Gather (1) quality assurance (QA) data and summary of findings for next-generation products, and (2) summary data from prototype fabrication and testing. The engineers in the product development department need to gather QA-related summary data from other departments, such as the production engineering and production operations departments, and must incorporate the needs indicated by these data into their product designs.

Table 2-13. Outline of Standard Procedure for Equipment Production Setup

Step-by-step management	Action to be taken	From whom	To whom	When
Preliminary evaluation (XQA investigation) (YQA investigation)	1 QA data questionnaire and needs for next-generation products	Engineering, etc.	Various dept heads	XYZ categories
	2 Needs from prototype and testing stages	Engineering, etc.	Various dept heads	XYZ categories
	3 Issue/receive production setup requests	From production management	Dept head	Immediately after receiving order
	4 Draft production setup schedule	Section head	Person responsible	Immediately
	5 Study final drawing (equipment modification)	Section head	Person responsible	Same day
	6 Check material shape needs	Section head	Person responsible	Immediately
Summary (six big losses) (ZQA investigation)	7 Preliminary evaluation of process planning	Production engineering	Various depts	Each time
	8 Draft part-specific process plan	Section head	Person responsible	Each time
	9 Set goals	Related depts	Production engineering	Each time
	10 Investigation and preliminary evaluation of fabrication methods	Production engineering	Various depts	Each time
	11 Fabrication method investigation meeting	Section head	Engineering, QA, manu-facturing management	Each time
	12 Determine process planning and fabrication methods	Section head	Person responsible	Each time
	13 Prepare to deliver equipment	Production engineering	Clerical	Each time

	14	Prepare to circulate approval memo	Person responsible	Dept and section heads	Each time
	15	Equipment decision meeting	People responsible in dept/section	Executive	Each time
I (Planning)	16	Preliminary evaluation of equipment	Production engineering	Related engineers	Each time
II (Drawing check)	17	Determining and distributing detailed specifications	People responsible in dept/section	Section head	Each time
	18	Plan equipment procurement schedule	Section head	Person responsible	According to main schedule
	19	Mid-term progress report	Dept and section heads	Section head	Twice per month
	20	Equipment completion check	Designers	Manufacturer	According to planning table
III (Fabrication and witnessed inspection)	21	Single test (trial A)	Design, fabrication, production operations	According to planning table	
IV (Installation)	22	Installation test (trial B)	Designers and fabricators	After all processes are installed	
	23	Guide operators in equipment operation and routine inspection	Designers and fabricators	Specified operators	After trial B
V (Commissioning)	24	Line test (trial C)	Operations dept	Related depts	Scheduled date for trial C
	25	Final debugging (trial D)	Operations dept	Related depts	After trial C
	26	Initiate commissioning control	Operations dept	Persons responsible on production line	From specified date
	27	Cancel commissioning control	Dept head	3 months after trial C	

Table 2-13. cont'd.

Step-by-step mgmt (cont'd)	How	Step-by-step output	Created by whom	Sent to whom	Follow-up action	Confirm-ation	Com-ments
Preliminary evaluation (XQA investi-gation) (YQA investi-gation)	1 Collect on QA data sheet	QA data sheet 1	Persons resp. in various depts	Requesting dept	Work out details of action plan		
	2 Request or provide fabrication method improvement to product designers	Finished equipment modification drawings	Engineering	Fabrication	Fabrication of modified equipment		
	3 Tell all members at morning meeting	Get and circulate request form			Content check by person responsible		Report to clerical dept
	4 Draft main schedule	Step-by-step mgmt table to prepare for equip. production	Section head	Dept head	Use for schedule management		
	5 Check by engineering dept	Receive final drawings	Receive final drawings	Person responsible	Exchange with old drawings		
	6 Mutual check by forge engineer and quality assurance manager	Material shape equipment modification request form	Person responsible	Engineering	Arranged between forge and quality assurance		
Summary (six big losses) (ZQA investi-gation)	7 Set orientation with related depts	Six big losses and QA data sheet 1	Shop floor	Production engineering	Categorize and summarize		
	8 Send orientation instructions to persons responsible	Part-specific process plan	Person responsible	Dept head	Carry out no. 10 fabrication setup		
	9 Set targets for production, quality, and costs	Create table of target values	Person responsible	Dept and sect. heads	Coordinate fabrication methods to targets		
	10 Detailed evaluation of fabrication methods based on step no.7	Create fabrication method evaluation table (DRFQA)	Person responsible	Dept and sect. heads	Preliminary eval. of new fabrication methods		
	11 Discuss and give notification of abnormalities from step no.7	Record in minutes of meeting	Person responsible	Dept and sect. heads	Incorporate into step no. 12		
	12 Discuss and give notification of abnormalities from step no.7	Process planning table and layout drawings	Person responsible	Dept and sect. heads	Confirm bottle-neck processes		
	13 Implement when there are items to be developed	———	———	———	———		

	14	Recheck contents of schedule	Approval memo request form	Person responsible	Supervisor	Departmental confirmation meeting		Send meet'g descript. to clerical dept
	15	Explain if over $75,000	Prepare documentation for OHP (put in minutes of meeting)	Person responsible	Section head	Supplemental explanation of instructions		
I (Planning)	16	Work out details centered on main equipment units	MP design case study (DRFQA)	Person responsible	Dept and section heads	Incorporate in equipment specifications		
II (Drawing check)	17	Work out M/C specifications	DRFQA (sheet 2)	Person responsible	Section head	Incorporate in equipment specifications		
	18	Work out details on each M/C	Detailed production setup schedule	Person responsible	Section head	Use this to check up to trial A		
	19	Check local manufacturers	Detailed production setup schedule	Various manufacturers	Section head	Progress check		
	20	Check specifications before trial A	Detailed production setup schedule	Various manufacturers	Section head	Judgment for trial A		
III (Fabrication and witnessed inspection)	21	5Ms + safety + quality check	Trial A report	Person responsible	Dept and section heads	Send proposals to operations depts		Send explanation to clerical depts
IV (Installation)	22	Carry out final debugging after trial A, then implement production engineering for test run	Complete and submit handover items	Production engineering	Manufact'g dept head	Transfer management from production engin'r'g to manufact'g		
	23	Training in routine inspection and operation of each machine (about 2 days)	Guidance report	Person responsible	Section head	Training drills run by manufacturing dept		
V (Commissioning)	24	Use line test outline (centered on operations dept)	Request for action on items pointed out by trial C	Manufacturing	Related depts	Specify via detailed improvement plans		
	25	Implement actions regarding abnormalities from trial C	Report on action on items pointed out by trial C	Manufacturing	Related depts	Setup for production start		
	26	Analyze any problems from initial operation	Initial operation problem memo	Manufacturing	Related depts	Production start		
	27	Complete actions on problems from initial operation	Request to cancel commissioning control	Manufacturing	Quality assurance	Transfer to regular management		Send report to clerical dept

Step-by-step management. The production engineers start at step 7 (preliminary investigation of process planning) in Table 2-13 using preliminary evaluation charts as the basis for their step-by-step management. They sort out problems related to the six big losses appearing in processing the mass-produced products (the new products and any highly similar products at this process) so that actions to correct those problems can be incorporated at step 8.

Goal-setting. At step 9 (setting goals), the part-specific process schedule is used as the basis for setting goals in three process-specific areas: production capacity, process quality, and process costs. All of this is done within the framework of the required product quality and product cost schedule.

Implementation of DRFQA. At step 10 (study and preliminary evaluation of fabrication methods) and step 16 (preliminary evaluation of equipment), you should carry out DRFQA (design review for quality assurance) to predict and resolve problems, and to pave the way for determining and distributing detailed specifications at step 17.

Goal-specific test runs. You must orient each test run toward a particular goal. At step 21, try out the equipment unit on its own before installing it in the production line to test out its basic functions. At step 22, (installation test), check that the equipment has been installed correctly. At step 24 (line test run), check the quality and smoothness of the installed equipment's operation as part of the production line. At step 25 (final adjustments), take care of any remaining bugs and check for stable operation.

Standard schedule. Standardize the procurement schedule for each equipment procurement (Figure 2-17). You should also standardize the daily schedule relating to each step in this process. This will guide your management of the minor repairs made at these steps (Figure 2-18).

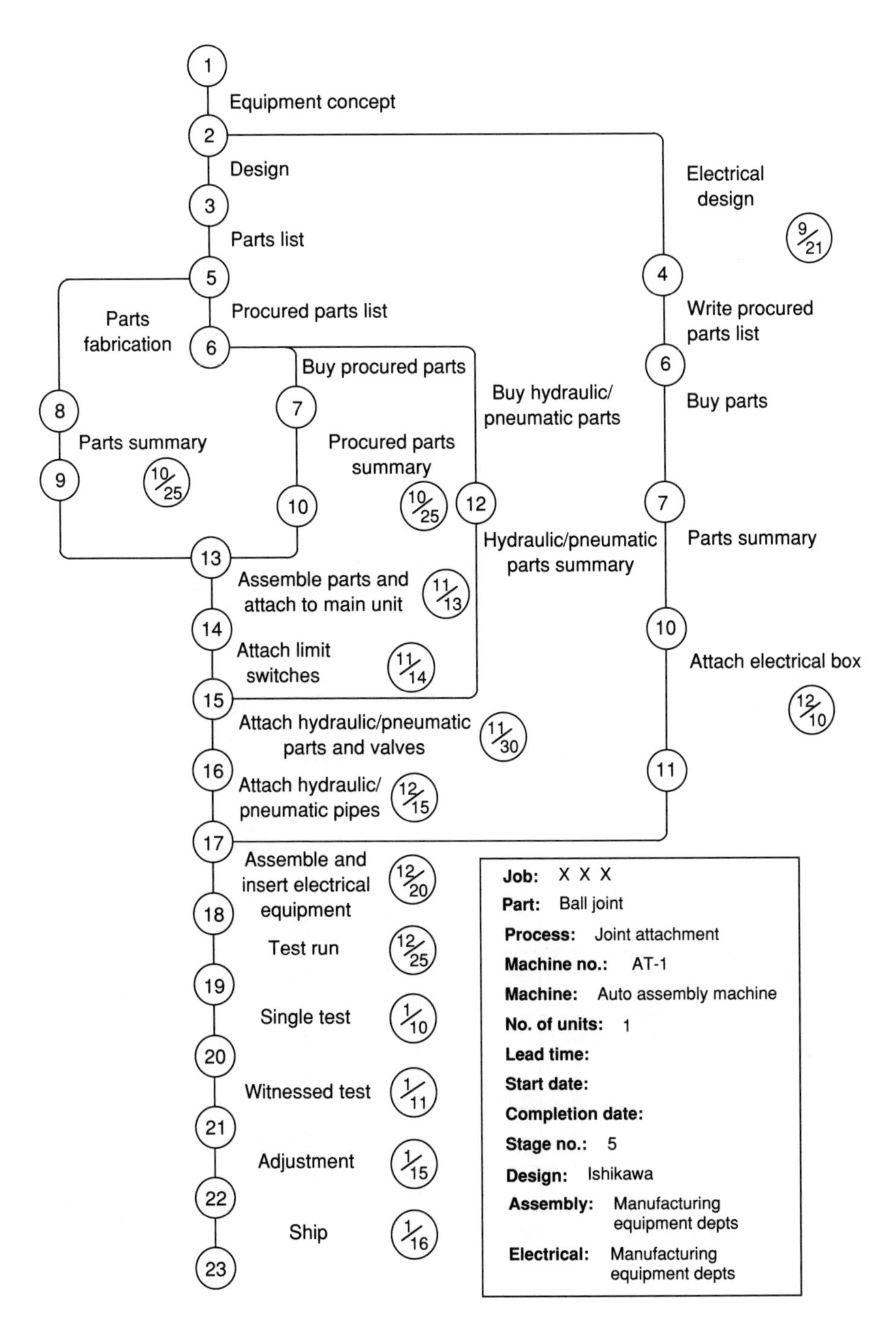

Figure 2-17. Part Diagram for Equipment Fabrication

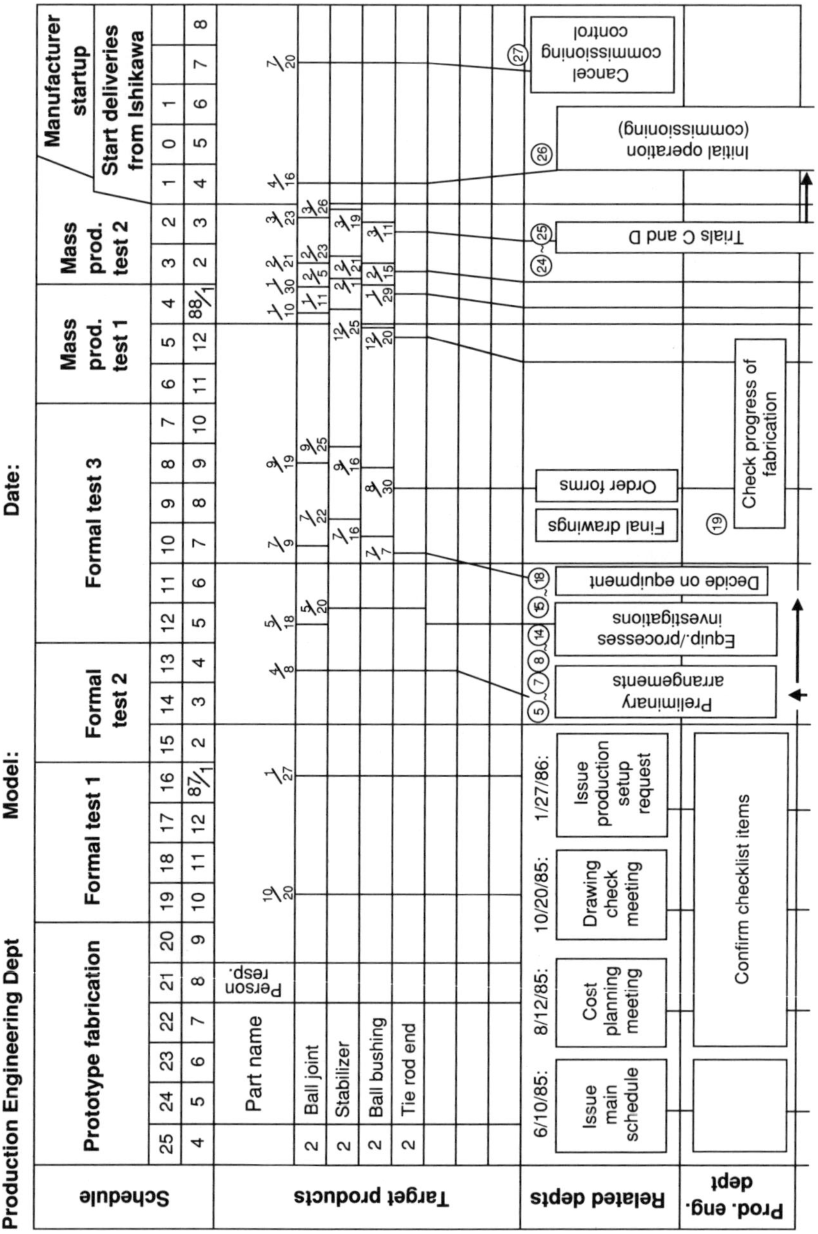

Production Engineering Dept **Model:** **Date:**

Schedule	Prototype fabrication						Formal test 1				Formal test 2			Formal test 3						Mass prod. test 1			Mass prod. test 2		Manufacturer startup / Start deliveries from Ishikawa				
	25	24	23	22	21	20	19	18	17	16	15	14	13	12	11	10	9	8	7	6	5	4	3	2	1	0	1		
	4	5	6	7	8	9	10	11	12	87/1	2	3	4	5	6	7	8	9	10	11	12	88/1	2	3	4	5	6	7	8

Target products	Part name	Person resp.	Dates
			10/20, 1/27, 4/8, 5/18, 7/9, 9/19, 1/10, 1/30, 2/21, 3/23, 4/16, 7/20
2	Ball joint		5/20, 7/22, 9/25, 1/11, 2/5, 2/23, 3/26
2	Stabilizer		7/16, 9/16, 12/25, 2/1, 2/21, 3/19
2	Ball bushing		7/7, 8/30, 12/20, 1/29, 2/15, 3/11
2	Tie rod end		

Related depts

- 6/10/85: Issue main schedule
- 8/12/85: Cost planning meeting
- 10/20/85: Drawing check meeting
- 1/27/86: Issue production setup request
- ⑤~⑦ Preliminary arrangements
- ⑧~⑭ Equip./processes investigations
- ⑮~⑱ Decide on equipment
- Final drawings
- Order forms
- ㉔~㉕ Trials C and D
- ㉖ Initial operation (commissioning)
- ㉗ Cancel commissioning control

Prod. eng. dept

- Confirm checklist items
- ⑲ Check progress of fabrication

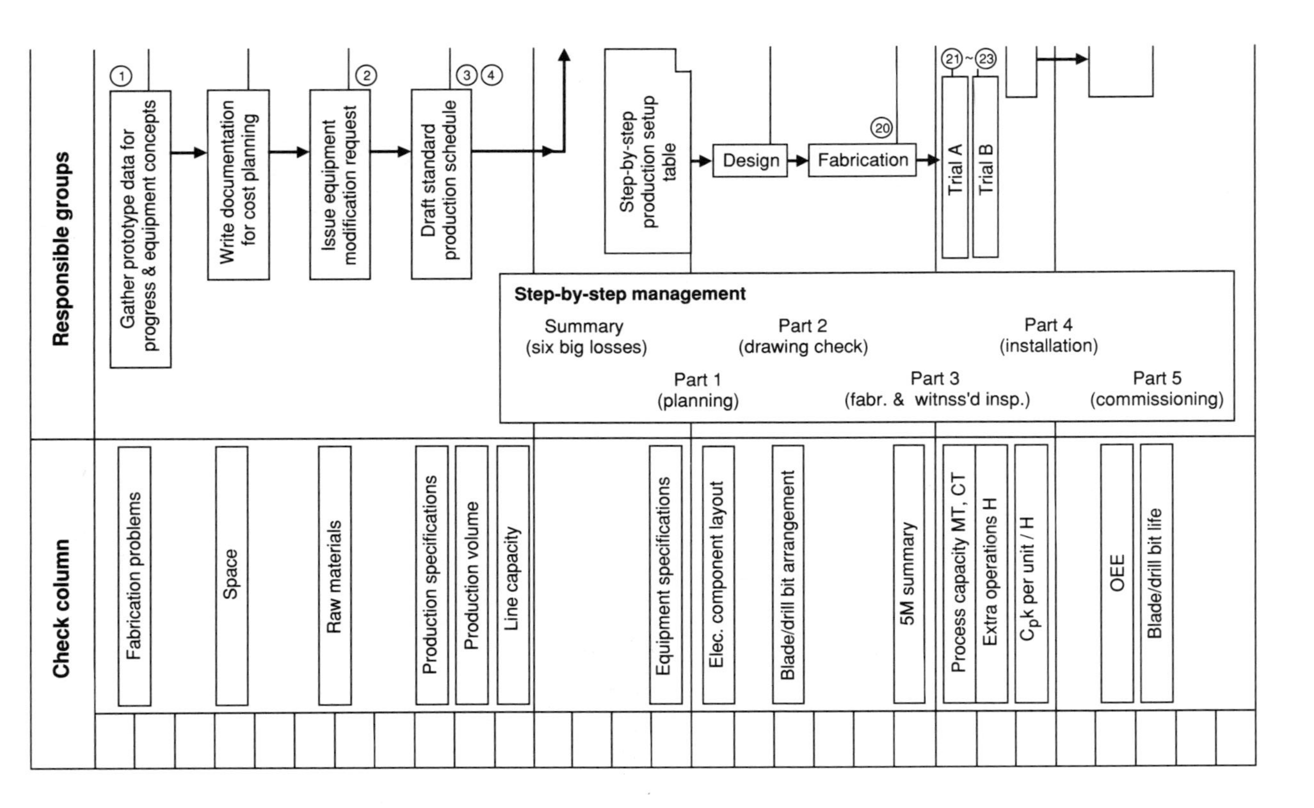

Figure 2-18. Step-by-Step Schedule Management Table for Equipment Production Setup

3

Quality Assurance (QA) Design

Whenever you take action against sporadic equipment breakdowns, it quickly becomes apparent that foresight at the equipment design stage could have prevented such breakdowns. This is also true for defects. In this era of automation, the quality control slogan "Build quality into products at each process" is more properly stated as "Build quality into the (automated) production equipment." In automated production, most defects are generated by the equipment, and in many cases the ultimate cause is lack of planning during the equipment design process.

Consequently, the secret for preventing defects lies in designing equipment that is easier to operate and maintain — in other words, "QA-friendly" equipment. What is QA-friendly equipment? To begin with, it is equipment that meets the following two requirements:

1. The equipment must include devices that can be trusted not to produce defects. (Hereafter, we shall call these conditions the "nondefective conditions.")

2. The equipment must be easily maintainable so that its nondefective conditions can be sustained throughout the equipment's life cycle.

EQUIPMENT RELIABILITY AND HUMAN-MACHINE SYSTEMS

QA-friendly equipment is necessary from the perspectives of equipment reliability and human-machine systems.

Put simply, equipment reliability is a measure of the likelihood that problems such as breakdowns and quality defects will not occur. Reliability can be broken down into intrinsic reliability and operational reliability, and therefore it is the sum of these two that is referred to as reliability or total reliability (see Figure 3-1). In this context, we shall consider reliability only as it relates to quality defects.

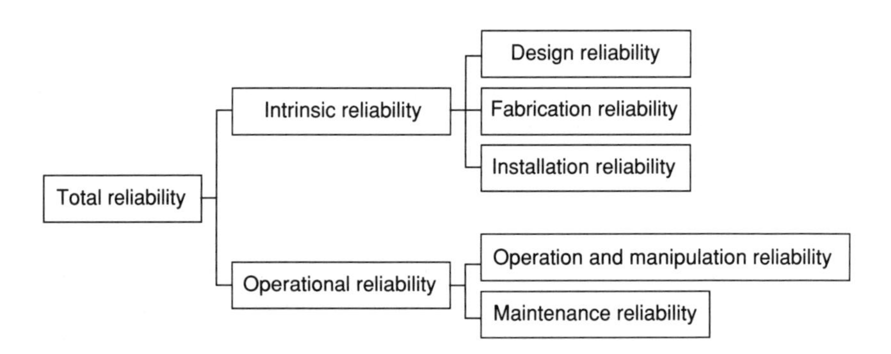

Figure 3-1. Equipment Reliability

Intrinsic reliability concerns strictly the equipment itself and is determined during the early equipment development stages by addressing the following two issues:

1. Does the equipment design ensure that the equipment has all the functions needed to meet all the conditions under the general concept of nondefective conditions? We call this kind of intrinsic reliability "design reliability."
2. Do all the equipment functions operate normally after the fabrication and installation stages? We call these kinds of intrinsic reliability "fabrication reliability" and "installation reliability."

Operational reliability is determined by how well the equipment's preestablished nondefective conditions can be maintained while the equipment is in use. Two factors help to determine this:

1. Can the equipment be operated as planned (operation and manipulation reliability)?
2. Can the equipment be restored to optimal condition from within a planned range of deterioration-related changes in its nondefective conditions (maintenance reliability)?

During the life of the equipment, there are three types of defects, which can be expressed in a bathtub curve diagram.

The first type of defect is the early defect, which occurs during the mass-production test-run stage and/or commissioning stage. Early defects that occur as a result of low intrinsic reliability include those caused by

- an inadequate or vague understanding of nondefective conditions,
- equipment that does not contain functions to fully establish nondefective conditions.

Early defects can also be due to low operational reliability, such as human errors caused by lack of training or experience in operation and maintenance of the equipment at the initial stages.

The second type of defect is the random or sporadic defect. These are usually due to human errors such as careless mistakes made during operation or maintenance, lack of training of new operators, or management errors such as supplying the wrong materials. In any case, these errors contribute to low operational reliability.

The third type of defect is caused by deterioration. As equipment deteriorates over time, its nondefective conditions can change without being noticed or restored, which again promotes low operational reliability.

Thus, when we speak of QA-friendly equipment, we mean equipment that has been designed so that all three types of defects are easily preventable.

EQUIPMENT FREE OF EARLY DEFECTS (INTRINSIC RELIABILITY)

The following concepts and approaches will help explain the causes of initial defects and reduce their occurrence.

How Does Intrinsic Reliability Decline?

How can equipment begin to have low intrinsic reliability? From the perspective of equipment reliability:

Equipment designers may lack technical expertise. They may have uncritically adopted an existing design (tendency to copy or adopt designs for novelty's sake); they may have left the design completely up to the equipment manufacturer, or failed to review detailed documentation because of time constraints; or the designers may have made errors in the design.

The equipment and the selected fabrication method may be ill-suited for the product design. It may contain requirements that are incompatible with the fabrication method and equipment under consideration, or vice versa.

From the perspective of fabrication and installation reliability:

Errors or omissions may have occurred during fabrication or installation. Some errors are due to lack of technical skill on the part of those carrying out fabrication or installation. In other cases, the designers may have left too much up to the discretion of people carrying out fabrication or installation and not pointed out all the important factors.

Debugging may not have been thorough during the test run. Although all of these major reasons are due to low operational reliability, they also contribute profoundly to poor intrinsic reliability.

Five Ways to Boost Intrinsic Reliability

With the above-mentioned perspectives on intrinsic reliability in mind, consider the following five ways to improve intrinsic reliability:

1. Work toward more effective product design.
2. Analyze equipment function and establish values to guarantee them.
3. Identify and respond early to problems in development of functional prototypes.
4. Confirm process capacity using CAD/CAM technologies.
5. Confirm process capacity and response to related problems at the mass-production prototype testing stage.

Work Toward More Effective Product Design

This method has already been described in Chapter 2, pages 35-57.

Analyze Equipment Functions and Establish Values to Guarantee Them

Before writing the equipment specifications manual, designers should not only establish specifications covering main items from process quality to equipment functions, they should also analyze and work out specification values that ensure the quality of every component factor.

Identify and Respond Early to Problems in Development of Functional Prototypes

Traditionally, only the functions or performance of the functional prototypes are tested to see if they provide all the required functions. Problems related to mass-production development are not anticipated and dealt with until that stage begins. As lead times have grown shorter, you may already have recognized that waiting until that stage only compounds any problems encountered.

A better approach is to surface and address mass-production problems at the functional prototype stage. This eliminates the need for a mass-production prototype — as long as you take preventive measures against problems that can occur after the mass-production stage. During the prototype stage you should identify problems related to mass-production development in the following ways as an additional emphasis in functional prototype testing:

Identify and address mass-production development problems through mass-production conditions at the functional prototype testing stage. Simulating mass-production conditions as much as possible during the functional prototype testing boosts your ability to predict problems that may arise during mass production. Mass-production conditions might include:

- processes and process sequences
- dies, jigs, and tools for equipment to be used

- operation methods, procedures, time and space requirements, and the like
- strict adherence to specifications (improvising with existing parts tends to obscure mass-production problems)

To assure thorough problem identification, include all mass-production process operators, maintenance staff, and engineering staff in the prototype evaluation process.

Develop and apply evaluation methods that reveal hidden problems in mass-production development. Even seemingly effective prototypes can lead to problems at the mass-production development stage. Developing and using thoughtful evaluation methods for small-lot prototype models (such as modification of evaluation conditions) can allow you to predict and address problems that would otherwise show up only during mass-production development.

Conduct a design review to discover mass-production development problems at the prototype evaluation stage. Draft a standardized design review checklist that emphasizes the search for mass-production development problems. Another way to find these types of problems is to involve highly skilled people in the evaluation process.

Final Follow-up Measures for Confirming Process Capacity

- Confirm process capacity using CAD/CAM technologies.
- Confirm process capacity and response to related problems at the mass-production prototype testing stage. You must not omit this essential, final follow-up measure.

QA-FRIENDLY (RELIABLE-OPERATION) EQUIPMENT

The following sections explain the decline of operational reliability and list approaches and procedures to ensure QA-friendly equipment.

Why Does Operational Reliability Decline?

Defects arising from poor intrinsic reliability are an obvious concern. Of even greater concern are problems that result in lower operational reliability. Operational reliability usually declines in one of the two following types of situations:

- When the operators and maintenance staff lack sufficient knowledge or skills,
- When the equipment itself has not been designed for high operational reliability.

From the perspective of equipment design, the latter situation presents a clear problem. Insufficient consideration of operational reliability during the equipment design process can have the following consequences:

- Failure to indicate optimal (precisely quantified) conditions clearly
- Expectations that experienced factory personnel will establish the nondefective conditions
- A specified range for the nondefective conditions that is either too narrow or too loose
- Operation procedures that are too difficult to carry out without errors and that are set simply for the sake of achieving nondefective conditions
- Changes in the nondefective conditions that go unnoticed until they result in defects or are ignored because the equipment is hard to understand
- Nondefective conditions changing so easily that troublesome fine-tuning or maintenance work is often required
- Deterioration-related changes in nondefective conditions that are hard to spot in the equipment or hard to restore

All the above cases occur quite frequently and lead to reduced operational reliability, sporadic defects, and chronic defects. Why do such cases occur so frequently? Equipment designers tend to see their work as centered on the equipment itself. This focuses their attention on functional design and intrinsic reliability. Rarely do they consider the equipment from the perspective of operators and maintenance staff, that is, in terms of human-machine system design (the relation between the equipment and the people who operate and maintain it).

Equipment-centered design is at the root of low operational reliability. The people who work with such equipment are required to adapt their work to suit the design, which is not always possible in practice even when it seems perfectly possible on paper.

Accordingly, equipment designers should consider equipment design quality as a factor of the human-machine system and approach design work as the task of establishing optimal relationships between human workers and machines. For example, even when defects are due to the operator's mistakes, the equipment designer should take the view that the equipment invites such operator errors. Human-machine system design requires you to look at the strengths and weaknesses of both the machine's and the human's roles and make designs compatible with both sets of characteristics.

Five Conditions for QA-Friendly Equipment

QA-friendly equipment must meet at least the following five conditions:

1. Clearly defined (quantified) nondefective conditions
2. Nondefective conditions that are easy to establish
3. Nondefective conditions that do not change easily
4. Changes in nondefective conditions that are easy to recognize
5. Nondefective conditions that are easy to restore

Establish Clearly Defined Nondefective Conditions

Figure 3-2 illustrates simply the relation between quality and equipment. The nondefective conditions are defined as

- processing conditions to meet process-specific quality standards that have been set to fulfill product design quality requirements
- equipment function and precision conditions required to support those processing conditions

Nondefective conditions can also be thought of as a set of factors, as shown in Figure 3-2.

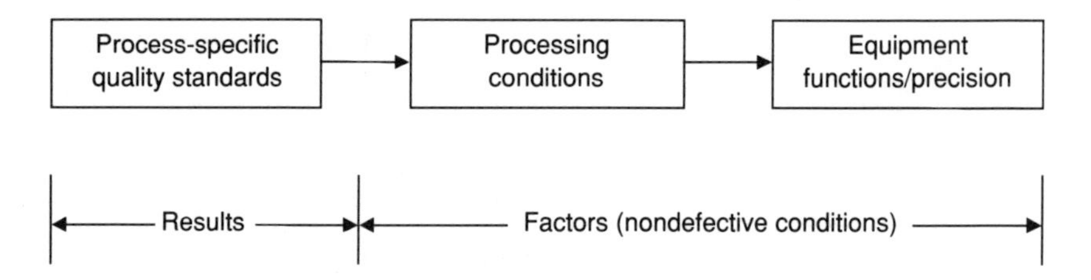

Figure 3-2. Relationship Between Quality and Equipment

To completely prevent defects, you must begin by clarifying all the factors behind the nondefective conditions. Quantifying all the nondefective conditions should be a part of the equipment design process. Sometimes, when working out the design, having quantified processing conditions is enough; sometimes it is useful to also quantify the equipment component conditions (e.g., friction tolerances) needed to maintain the processing conditions at a stable level. When a defect occurs, the problem often lies in the equipment component conditions.

All too often, it is hard to understand how to read the characteristics of the nondefective conditions. Even if you do understand how to read them, you may not know how to quantify them or set clear levels and tolerance ranges for them.

Consequently, when defects occur, you may have trouble pinning down the causes.

At a machining process, for instance, a quality team may have identified vibration in the grinder's main spindle or grindstone or the condition of the machine's coolant as a factor contributing to the defect. If the team hasn't specified the tolerances quantitatively, however, it cannot restore and maintain the grinder's nondefective conditions. Likewise, at the painting process, the team may already know that the booth's air-conditioning balance has a significant impact on the painting quality, but unless it ascertains clearly and precisely the characteristic values of the proper balance (admittedly, finding those values may involve difficult measuring problems), it cannot know exactly what conditions are required to maintain a stable air-conditioning balance for the painting booth.

At an arc-welding process, causes of welding defects are usually understood to include factors such as precise welding point location (workpiece precision and the arc-welding device's mechanical precision), welding direction, angle of approach and retreat, and the welding feed rate, or welding speed. Nonetheless, the tolerance and limit values still remain unclear.

At the factory, people judge the process conditions by the results (workmanship in the product, bead shapes at welding points, etc.) and try to establish the correct conditions by trial and error. Since this method is not reproducible, it is difficult to prevent defects completely.

Make Nondefective Conditions Easy to Establish

Even when nondefective conditions are perfectly clear, they may still be difficult to establish or may be prone to operator errors. Continuing with the painting booth example, suppose that the team already knows the precise values required for air temperature and humidity inside the booth. Other problems may still exist. Perhaps the sensors are not accurate enough or are set in the wrong places. Perhaps the booth's control system itself is inadequate and responds too slowly to condition changes. Similarly, an injection molding machine may be hard to accurately set and adjust because the dials that set the injection time and injection pressure have markings that are imprecise or are too small to see easily.

Make Nondefective Conditions Resistant to Change

If nondefective conditions are not stable after being set, it becomes difficult to maintain those conditions and ensure the proper quality levels. At a welding process, there are reference surfaces or pins for the workpiece positioning jig. If these are prone to rapid wear or electrolytic corrosion, or if they tend to collect sputtered welding debris, the jig's precision will worsen steadily. Likewise, processes that employ industrial robots are susceptible to occasional operational errors caused by electromagnetic noise. In fact, electrical and hydraulic devices in general are susceptible to interference from similar neighboring devices when the devices are lined up barracks-style. This can make nondefective conditions unstable, hard to maintain, and more likely to produce defects.

Make Changes in Nondefective Conditions Easy to Recognize

Make any changes in the nondefective conditions obvious enough so that anyone can recognize them and take preventive action. Since nondefective conditions consist of a group of factors, being able to recognize changes in those factors lets you know when defects are likely to occur (if the changes are allowed to continue) and to act before the defects occur.

In the earlier example of a jig with reference planes or reference pins that are gradually worn down, the wear tolerances may already have been precisely quantified. In some cases, however, it takes too much time or trouble to measure the current wear conditions using a caliper or micrometer, and so excessive wear conditions are not noticed until they result in a conspicuous number of defects. In one case, for example, a team tried various measures to solve a stubborn trend of increasing defects, only to discover finally that its inspection jig had been wearing down so that more and more perfectly good products were being judged as defective! In this case the inspection jig's design made it hard to measure precisely.

In factories that use general-purpose equipment such as injection molding machines, maintenance personnel should monitor the wear conditions of the machines' check rings and screw bind. If there are too many injection machines to justify so many precise measurements, they may check the results (number of defects produced) instead. The key to facilitating preventive action in this case, however, is to design injection machines with wear conditions that are easy to measure quickly. This enables the maintenance staff to monitor the factors instead of the results.

Make Changes in the Equipment's Nondefective Conditions Easy to Restore

Even when it is easy to detect changes in the nondefective conditions of a particular machine at an early stage (when preventive action is possible), restoring the machine's nondefective conditions may still require too much time, labor, or expense. Consequently, the equipment designers must also consider the machine's maintainability whenever deterioration is a problem or risk. This concern should be extended to all function-related parts that directly influence processing conditions or quality. You may have noted already the importance of making equipment less susceptible to deterioration, but if such durability comes at a high cost, it may be better instead to design the machine so that wear conditions can be monitored easily and worn parts replaced easily (such as by using the block exchange method).

Some machines, such as those used at painting processes, contain pumps, blowers, or other rotary devices that break down frequently and have a big impact on nondefective conditions. After estimating the impact of such devices on breakdown, quality, and cost conditions, you may find that the most secure and cost-effective measure is to design for redundancy so that malfunctioning devices can be instantly replaced by backup devices.

In addition to these five conditions for QA-friendly equipment to be considered at the design stage, it is also important at the operational level to analyze thoroughly the causes of defects that have already occurred to find clues for defect-preventive design measures. In other words, equipment designers must not shy away from going to the production line to find and employ data in support of QA-friendly equipment design.

Case Study 3-1: QA-Friendly Equipment (Toyota Auto Body)

Consider the following example of QA-friendly equipment design developed at an auto body plant.

Problems with Conventional Equipment Design

This case study demonstrates how a failure to consider QA-friendly design at the design stage caused problems for Toyota Auto Body at its auto body assembly processes. Specifically, these problems concerned the sliding side door on commercial van bodies that were coming off the line some years ago. The closed doors were sometimes not completely flush with the side of the van. This imprecision in slide-door positioning is an important factor in terms of the van body's external appearance, but it remained difficult to ensure precision.

Numbers 1 through 6 in Figure 3-3 show the precision measurement points. Figure 3-4 shows how points 3 and 4 tended to change over time. The latter figure indicates that by the mass-production startup stage QA check, precision in slide-door positioning had gradually deteriorated. The ultimate design-related cause was failure to make the factory equipment's non-defective conditions easier to maintain.

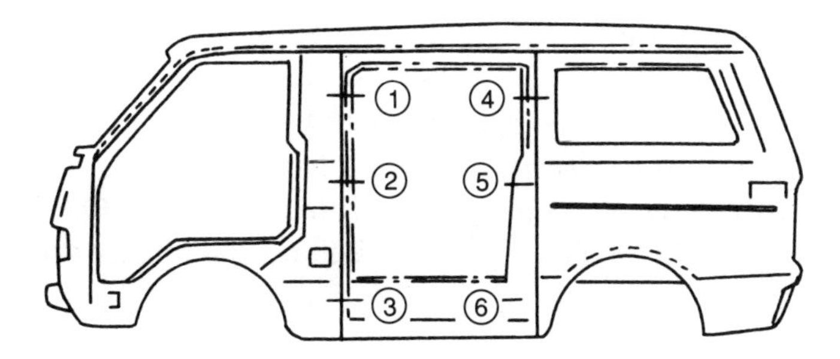

Figure 3-3. Precision Measurement Points

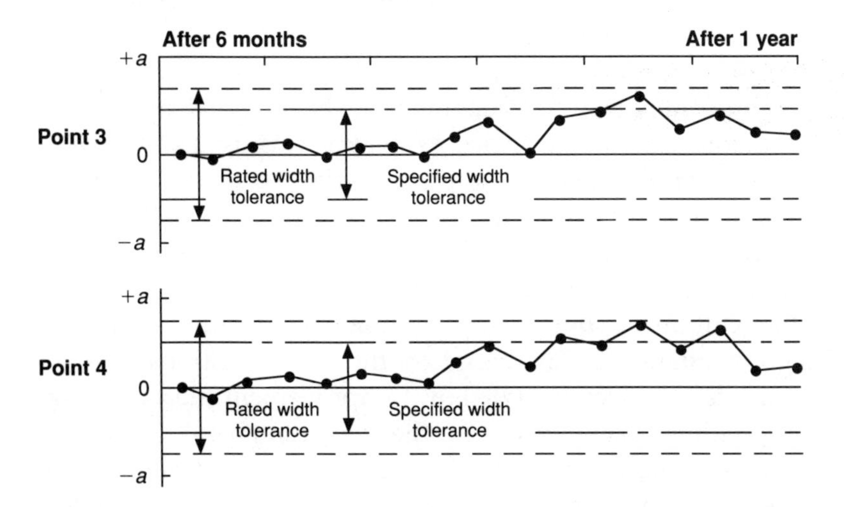

Figure 3-4. Precision Changes at Two Points after Production Startup

The fabrication method development chart in Figure 3-5 shows the role of the member subassembly jig. In this example, the wear in the support surface for the feed lifting pressure bar and play in the pressing surface bearings are directly responsible for problems. The operation manual indicated that this bar should be inspected for wear damage once a year using a height gauge, and that the compensation level (wear tolerance range) is 0.3 mm. Examining these conditions in light of the five conditions for QA-friendly design reveals, however, that the equipment's conditions change easily (the equipment wears down easily), and that such changes are both hard to recognize (height gauge is hard to use accurately) and hard to restore (parts are not easy to replace). Thus, the equipment is more QA-hostile than QA-friendly, which is a factor behind the results shown in Figure 3-4.

Design Review Method for Auto Body Processes

This example shows a design review at Toyota Auto Body from the standpoint of QA-friendly equipment design.

Figure 3-6 shows part of a fabrication method development chart concerning the construction of the back door, a main quality component in new vehicles. When this was used as a preliminary evaluation chart, the sections marked as *(B)*, *(C)*, and *(D)* were added to help flush out deterioration-related defect problems and make preventive improvements. Section (B_1) contains column headings based on the process management conditions for the equipment components; section (C_1) contains equipment design evaluation points based on the five conditions for QA-friendly design; section *(D)* shows improvement measures and corresponding cost estimates for the evaluation results.

Armed with these tools, the design team used the preliminary evaluation table to reveal hidden deterioration problems and to plan several improvement measures, each of which involved a cost analysis to find the most cost-efficient improvement measure. Figures 3-7 and 3-8 show the improvements that were made in this case.

Using the equipment evaluation method just described, the design team realized the following advantages:

1. They were able to clarify at the equipment design stage abnormalities that could lead to quality problems during routine production.
2. They were also able at that stage to make cost estimates of improvement proposals to address the quality problems.

QA DESIGN FOR INTRINSIC RELIABILITY AND OPERATIONAL RELIABILITY

Although intrinsic reliability and operational reliability were discussed separately earlier, the ultimate purpose of QA design is to create equipment that serves both types of reliability at the same time. Accordingly, this section discusses the application of QA design in the preliminary evaluations of equipment and of fabrication methods for ensuring both types of reliability.

Accurate and Sustainable Operations and QA Design

In the following example, a company is studying ways to automate a hitherto largely manual assembly process, in response to needs for modification of production specifications and a higher level of process quality. Usually, the most important factor in automating assembly work is the accuracy and continuity of assembly motions. For example, in manual work, accuracy sometimes suffers. This is not due to any time-related trends but rather to occasional (random) operator errors, especially when the work includes blind operations or operations that depend on the operator's memory of the correct procedures. When the operator is working in a tight space, you may find more instances of omitted parts. Manual operations can

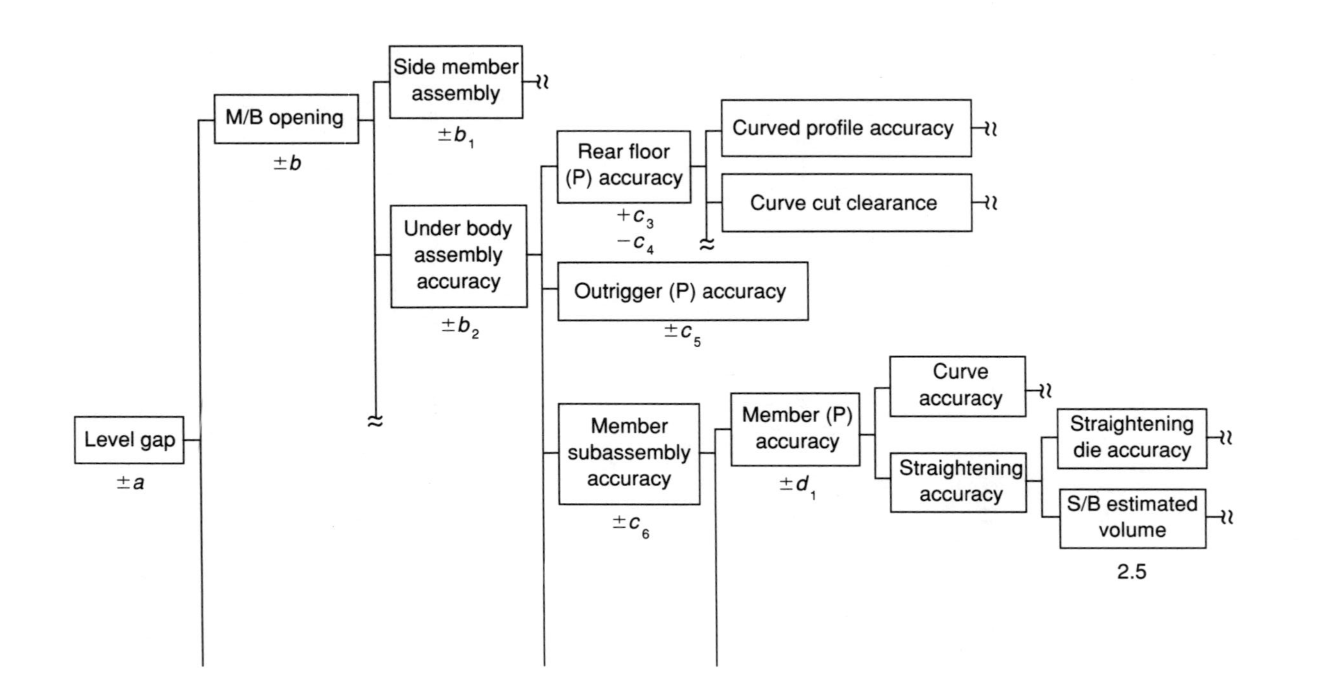
Level gap
±a
M/B opening
±b
Side member assembly
±b_1
Under body assembly accuracy
±b_2
Rear floor (P) accuracy
+c_3
−c_4
Curved profile accuracy
Curve cut clearance
Outrigger (P) accuracy
±c_5
Member subassembly accuracy
±c_6
Member (P) accuracy
±d_1
Curve accuracy
Straightening accuracy
Straightening die accuracy
S/B estimated volume
2.5

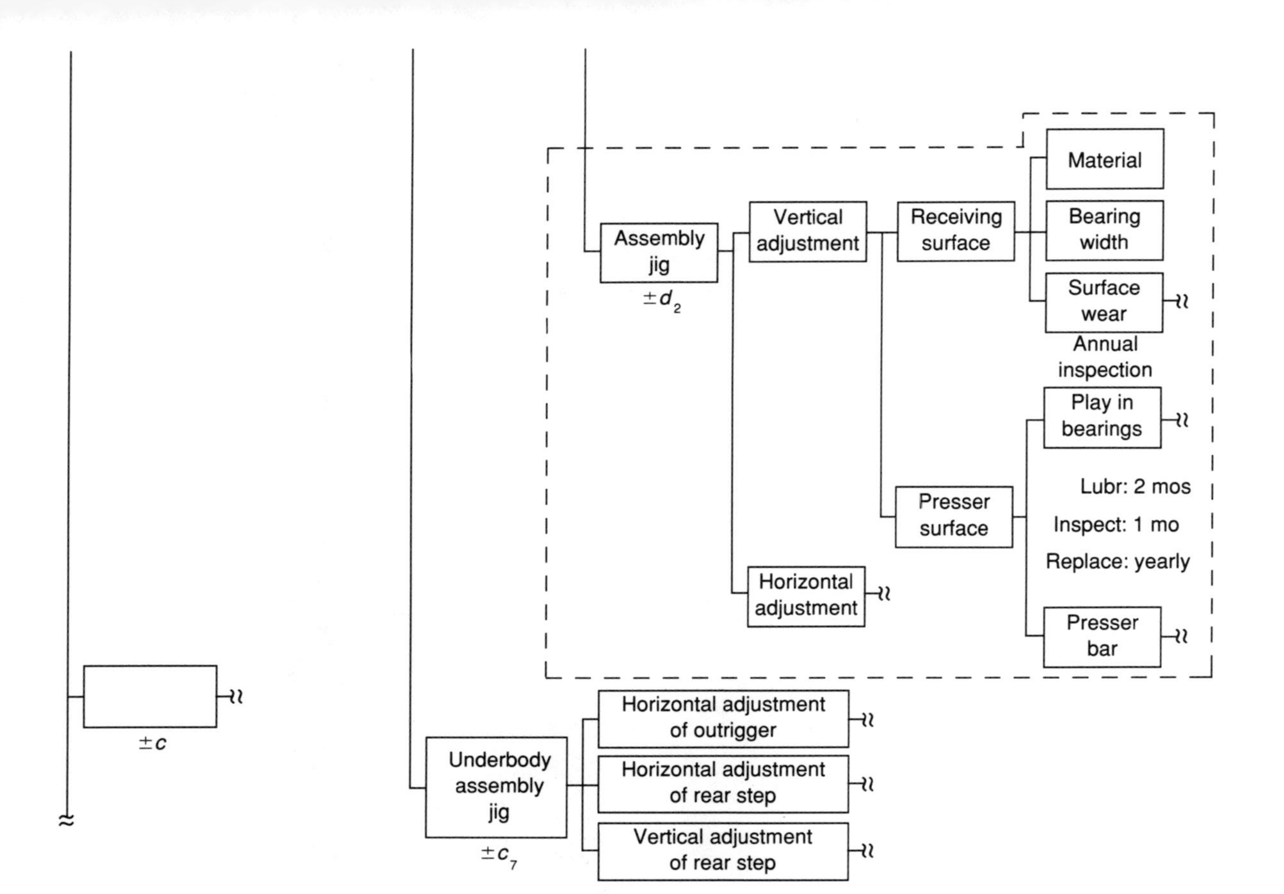

Figure 3-5. Fabrication Method Development Chart

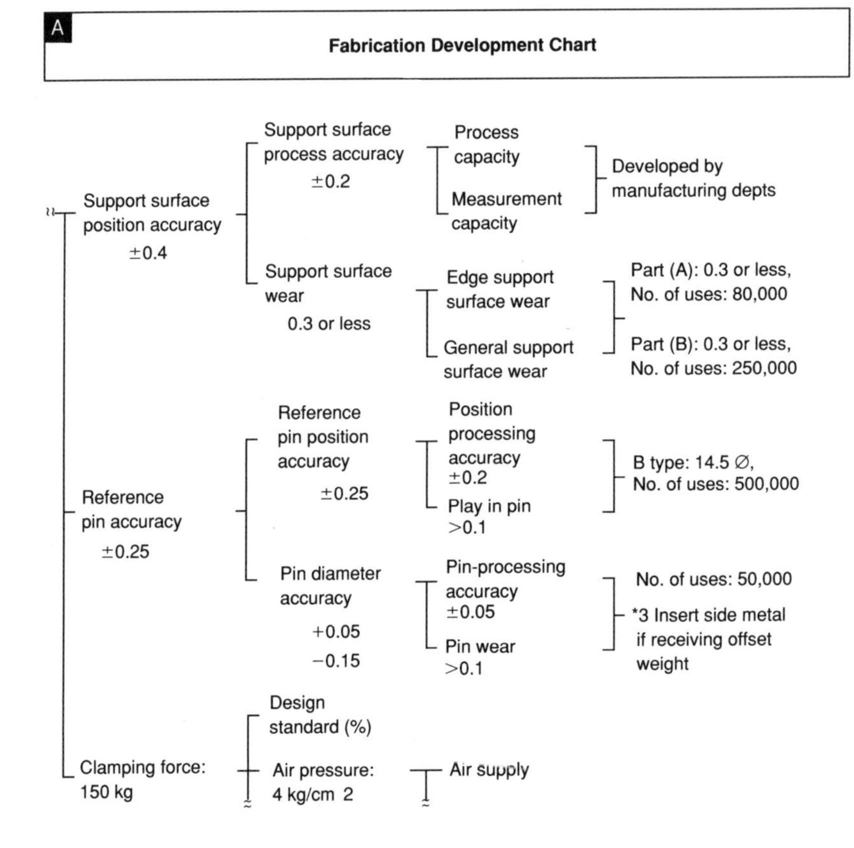

Figure 3-6. Fabrication Development Chart (New Vehicle Back Door Construction Quality)

B Clarifies process control items described in (A)				C QA-friendliness evaluation of each item from (B)					D Cost evaluation of x'd item from (C)	
				Five QA-friendliness conditions						
Control item	Inspect	Compensation	Tool	(1) Clarify	(2)	(3)	(4)	(5)	Improvement item	Cost
Wear	2/yr	0.3	Visual	○	○	×	×	○	----	---
"	1/yr	"	"	○	○	×	×	○	----	---
					See Figure 3-7	→ ○	○ ←	See Figure 3-8	Reevaluation	
Settling in pin mounting seat	1/yr	0.1	Touch	○	○	○	○	○		
Amount of wear	2/yr	Pin replacement	Calipers	○	○	○	Δ	○		
Factory air press	1/day	5.3 kg/cm or more	Press gauge	○	○	○	○	○		

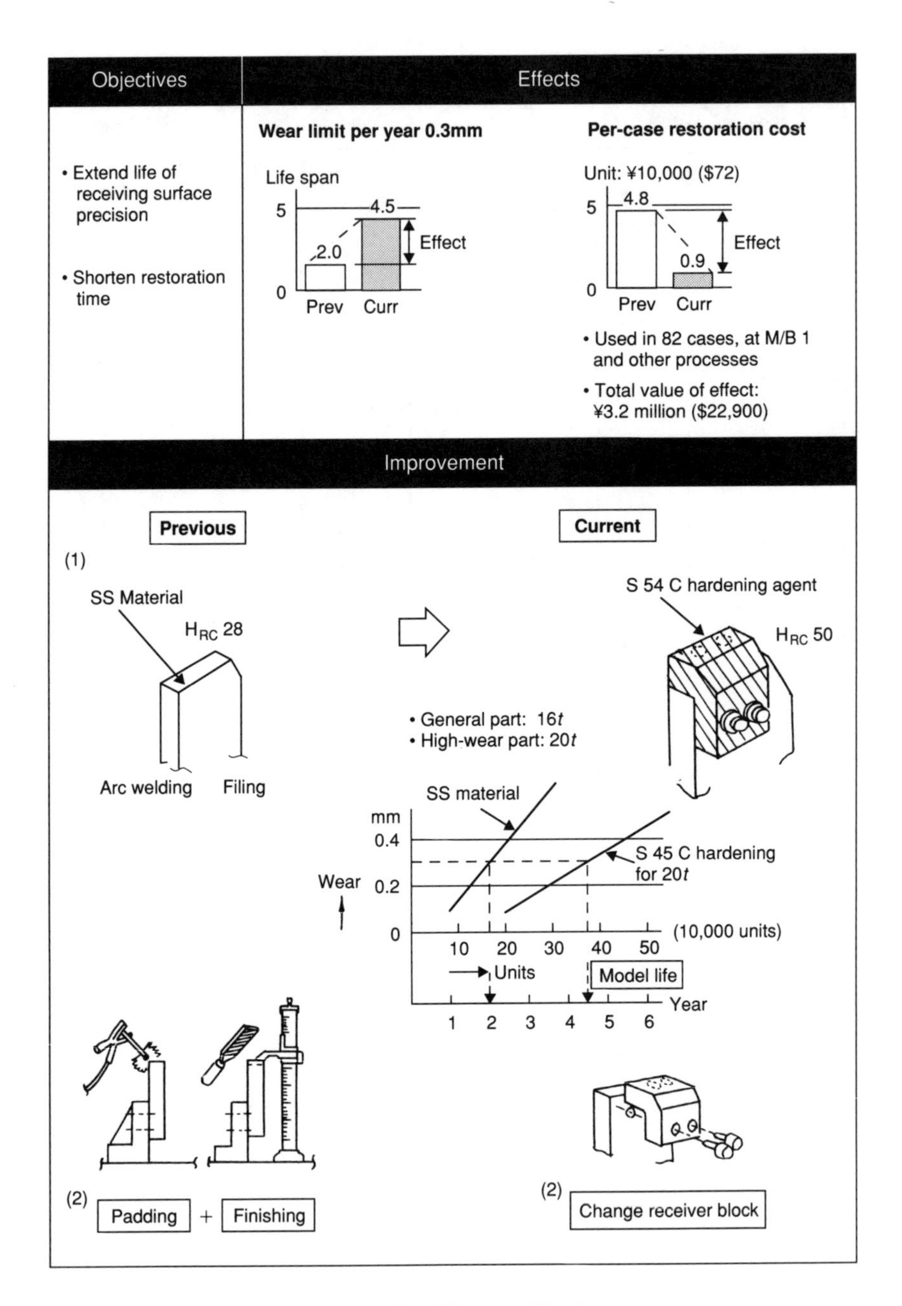

Figure 3-7. Difficult to Change but Easy to Restore

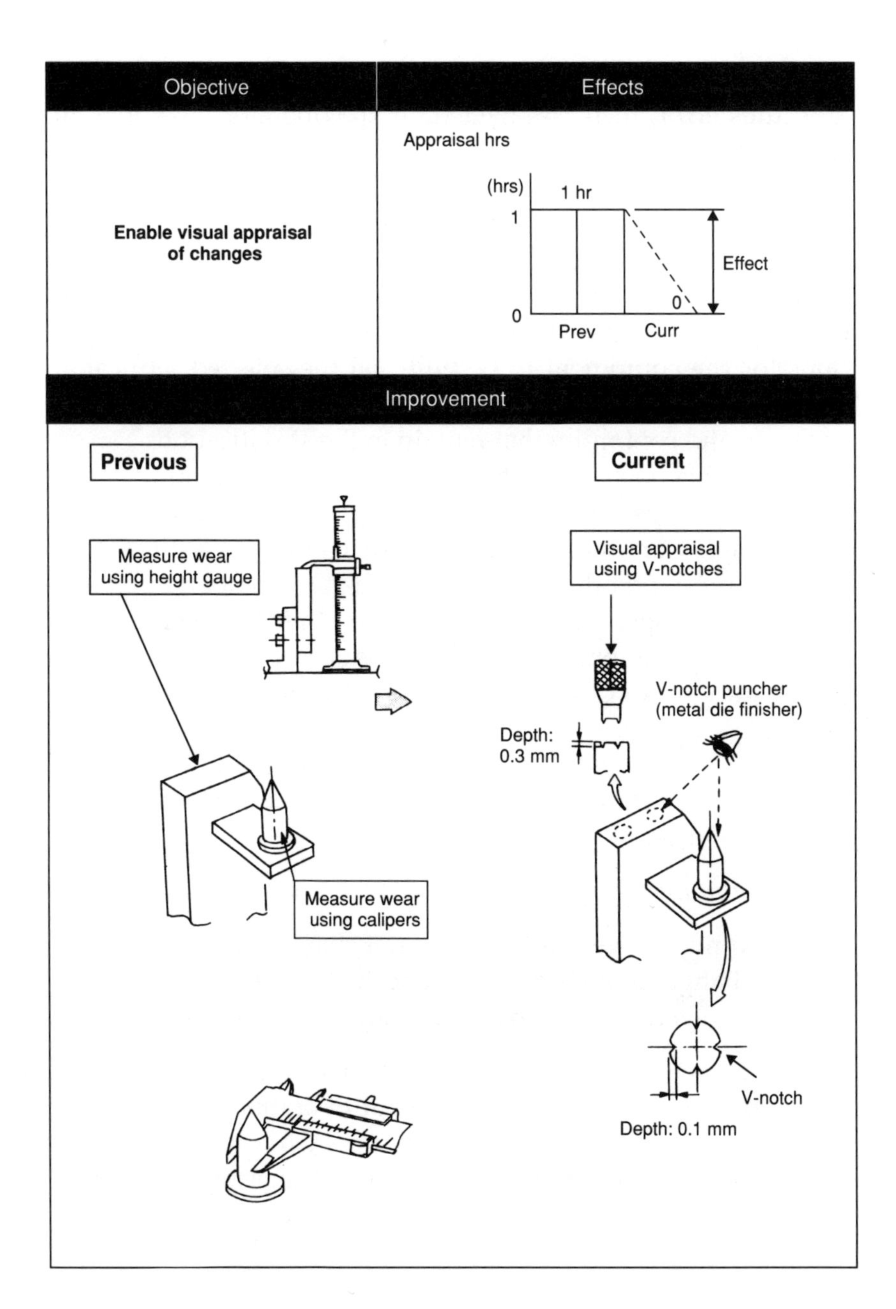

Figure 3-8. Improved Methods for Measuring Changes

also be susceptible to poor continuity when operator fatigue (and thus error) increases over time, the operator must look up often when doing the work, or when the tools wear down rapidly beyond the tolerance level.

Consequently, when automating, identify all the obstacles to accuracy and continuity that exist within the manual operations and find ways to build machines that take over these error-prone functions. Next, take another look at the specifications for the equipment to be built via the selected fabrication methods and identify and remove any obstacles to operational accuracy and continuity that remain in the specifications.

Accuracy factors in equipment operations include conditions such as operational precision, pressure, flow rate, current, and voltage. Continuity factors include conditions such as wear, deterioration, and other time-related changes in operating precision or other operating conditions. Whether for operations or equipment operations, these conditions can be matched with the five conditions for QA-friendliness, as shown below.

Accuracy:

- Nondefective conditions are precisely quantified.
- Nondefective conditions are easy to set.

Continuity:

- Nondefective conditions do not change easily.
- Changes in nondefective conditions are easy to see.
- Changed nondefective conditions are easy to restore.

Although you can apply the five conditions for QA- friendliness to both manual and equipment operations, you need to break down all operations into each type. Observe how this has been done in the following example of automating an assembly process.

Case Study 3-2: QA Design for Intrinsic Reliability and Operational Reliability

Equipment designers have traditionally tried to build equipment that will maintain consistent quality and operational reliability when new products are started up during regular production. Quality problems still arise, however, because the conventional equipment design is ill prepared to deal with the many modifications and adjustments that must be made manually when new products are introduced. This is precisely why you should use the five conditions for QA-friendly equipment as the basis for equipment design activities: to ensure higher levels of operational accuracy and continuity.

In the following case study such equipment design activities sought to improve upon manual operations by doing three things: (1) anticipating and addressing problems while determining the fabrication conditions, (2) anticipating and addressing problems while working out an equipment design that incorporates the above fabrication conditions; and (3) carrying out step-by-step management at every stage, from planning to regular production to ensure the success of measures taken at previous stages.

Comprehensive Preliminary Evaluation of Fabrication Methods

Comprehensive preliminary evaluations are commonly used in establishing fabrication methods. However, these conventional methods generally depend on specialized operation efficiency evaluation sheets and special skills on the part of the process designer. As such, these methods fail to find

- the specific problems related to variation in manual operations

- all the essential conditions (functions) that are needed to resolve the quality problems inherent in the manual operations

These activities can be more successful if a modified preliminary evaluation chart such as the one shown in Figure 3-9 is developed and used to resolve the various problems that crop up when introducing a new product. This chart is for preliminary evaluation of fabrication methods for assembling and mounting front differentials, which are key mechanisms in four-wheel drive vehicles. Section *(a)* includes basic assembly tasks that are assumed to be manual operations. Section *(b)* shows the fabrication methods and the evaluation items. This 19-point breakdown of the assembly factors deemed necessary for establishing the five conditions for QA-friendly equipment helps prevent omission of any condition that needs to be incorporated into the equipment. Each planned operation should be checked against these 19 evaluation points.

Section *(c)* describes specifically any problems emerging as a result of the preliminary evaluation in section *(b)*. Then, it lists in the "needed functions" column all functions needed to resolve these problems. Section *(d)* contains a menu of available fabrication methods from which to select the one best suited for automating the maximum number of problematic manual operations.

Following this type of procedure from conceptualization clarifies at the planning stage various types of problems (such as those concerned with product design, fabrication method selection, or equipment design).

Comprehensive Preliminary Evaluation of Equipment

After completing preliminary evaluation of fabrication methods and establishing the fabrication method conditions, the design team can begin translating those conditions into the

equipment design. At that time they should address in the equipment design the various QA issues that were uncovered in the course of establishing the fabrication method conditions. The first step in doing this is to implement a preliminary evaluation for the equipment.

The team can use the same fabrication method development chart shown in Figure 3-6 for this purpose. Figure 3-10 shows how this chart looks when used to evaluate equipment for an automatic bolt-fastening machine that assembles newly introduced front differentials. This chart differs from that shown in Figure 3-6, which was concerned mainly with equipment functions in a rear-door assembly process. Looking at the characteristics of the assembly equipment in terms of the five conditions for QA-friendliness, these conditions are broken down into 17 evaluation items that help ensure easy-to-use equipment.

Step-by-Step Management of QA-Friendly Equipment Design

In figures 3-9 and 3-10, all items that are marked with a triangle are continuing QA-related concerns. Even though certain concerns may be addressed by improvements during the course of the preliminary evaluation, they remain concerns until the equipment design has taken final form and the preliminary evaluation process is complete. Consequently, the team should apply thorough step-by-step management to this process of working out fabrication methods and equipment conditions to ensure quality at every step, from new product startup to regular production. An example of the step-by-step management chart appears in Figure 3-11.

Using these charts, equipment teams can list in detail the various issues to be evaluated at each step of the production setup process and reevaluate these issues from the standpoint of the five conditions for QA-friendly equipment. The teams can identify and follow up on new concerns at each step and

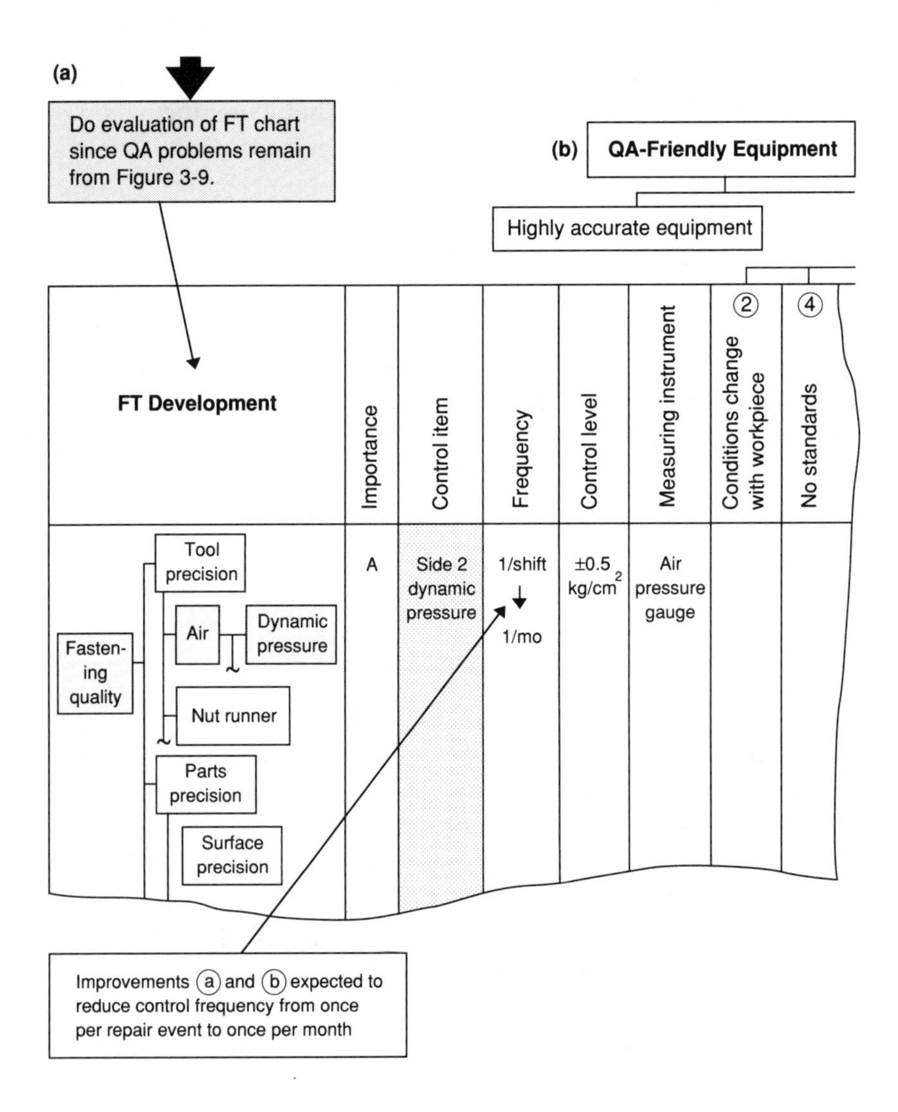

Figure 3-10. Equipment Preliminary Evaluation Form for Automatic Bolt-Fastening Machine (Front Differential-New Vehicle Model)

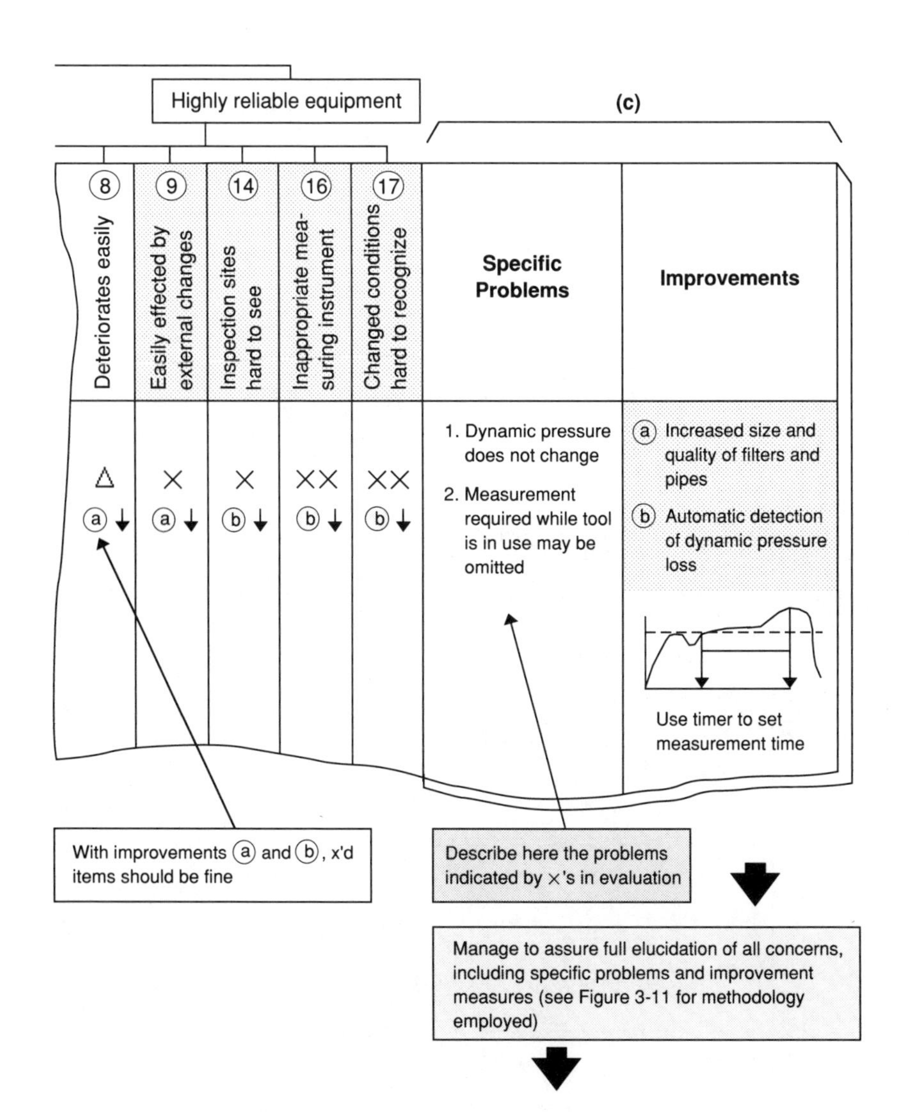
Highly reliable equipment
(c)
8
Deteriorates easily
9
Easily effected by external changes
14
Inspection sites hard to see
16
Inappropriate measuring instrument
17
Changed conditions hard to recognize
Specific Problems
Improvements
△
×
×
××
××
a ↓
a ↓
b ↓
b ↓
b ↓
1. Dynamic pressure does not change
2. Measurement required while tool is in use may be omitted
a Increased size and quality of filters and pipes
b Automatic detection of dynamic pressure loss
Use timer to set measurement time
With improvements a and b, x'd items should be fine
Describe here the problems indicated by ×'s in evaluation
Manage to assure full elucidation of all concerns, including specific problems and improvement measures (see Figure 3-11 for methodology employed)

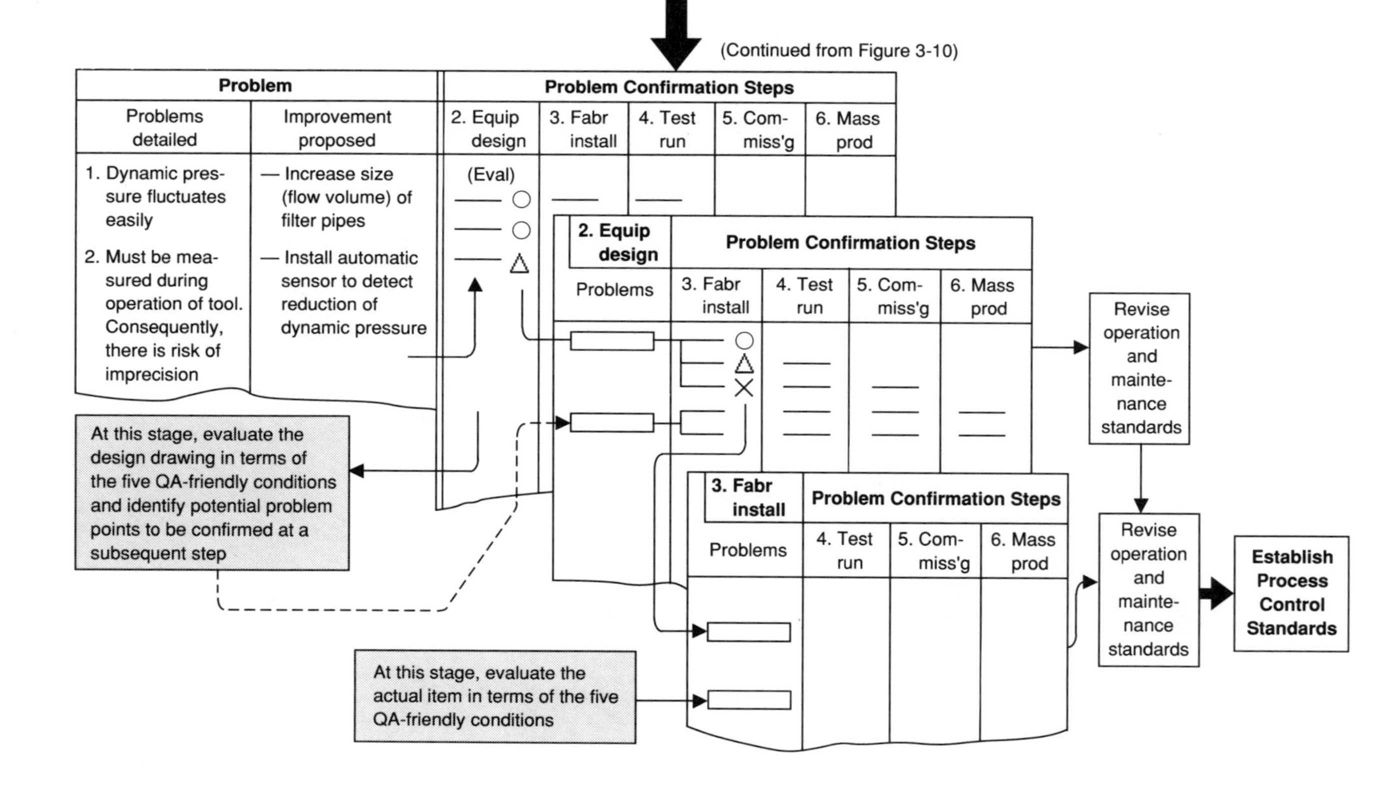

Figure 3-11. Step-by-Step Management Chart

resolve almost all problems having to do with either the fabrication methods or the equipment before the new products are started. The charts themselves describe the specific procedures.

MAINTENANCE QUALITY (MQ) MANAGEMENT AND QA DESIGN

This chapter began with a discussion of the QA design approach, and it concludes with a brief examination of QA design's position within total productive maintenance (TPM). The activities related to the quality aspect of TPM are called MQ (machine quality) control or quality maintenance. Figure 3-12 shows where quality maintenance fits in the context of the six fundamental TPM development activities. Table 3-1 matches up these six TPM activities with the main themes in MQ control. Finally, Figure 3-13 maps out the entire MQ control development process.

Fundamental TPM activities / TPM output	Equipment improvement	Planned maintenance	Autonomous maintenance	Early equipment management	Factory-friendly equipment or product design	Education and training
P: Production						
Q: Quality				QA design		
C: Cost						
D: Delivery						
S: Safety						
M: Morale						

MQ control themes

Figure 3-12. MQ Control Themes

Table 3-1. TPM Development Activities and MQ Control Themes

Six TPM Activities	MQ Control Activities
1. Equipment improvement	Carry out equipment improvements related to defect production (using P-M analysis)
2. Planned maintenance	Use diagnostic techniques and maintain equipment precision to assure quality (defect prevention)
3. Autonomous maintenance	Equipment-centered factory: training for engineers and technicians Labor-centered factory: shopfloor skill-building activities for quality assurance
4. Early equipment management	QA design: • Design equipment for zero defects from startup (intrinsic reliability) • Design QA-friendly equipment (operational reliability)
5. Factory-friendly product design	Design products to assure quality
6. Education and training	Teach engineers and technicians how to carry out M-Q activities

4
LCC Design

LCC, or life cycle cost, refers to the total cost of equipment (or an equipment system) throughout its life. The U.S. Office of Management and Budget defines LCC as "the sum of the direct, indirect, recurring, non-recurring, and other related costs of a large-scale system during its period of effectiveness. It is the total of all costs generated or forecast to be generated during the design, development, production, operation, maintenance and support processes."* In terms of production equipment, LCC can be described more simply as design and fabrication costs (initial or acquisition costs) plus the operation and maintenance costs (running or sustaining costs) In design-to-cost (DTC), we consider the LCC of a system (or equipment unit) as one of the design factors, along with other design specifications such as precision, speed, volume, weight, reliability, maintainability, and so on. The cost is not considered a result of the design, but rather a system target. Once we determine the entire system's target cost at the preliminary development stage, we distribute those

---

* U. S. Office of Management and Budget. 1976. *Major Systems Acquisitions.* Circular #A-109.

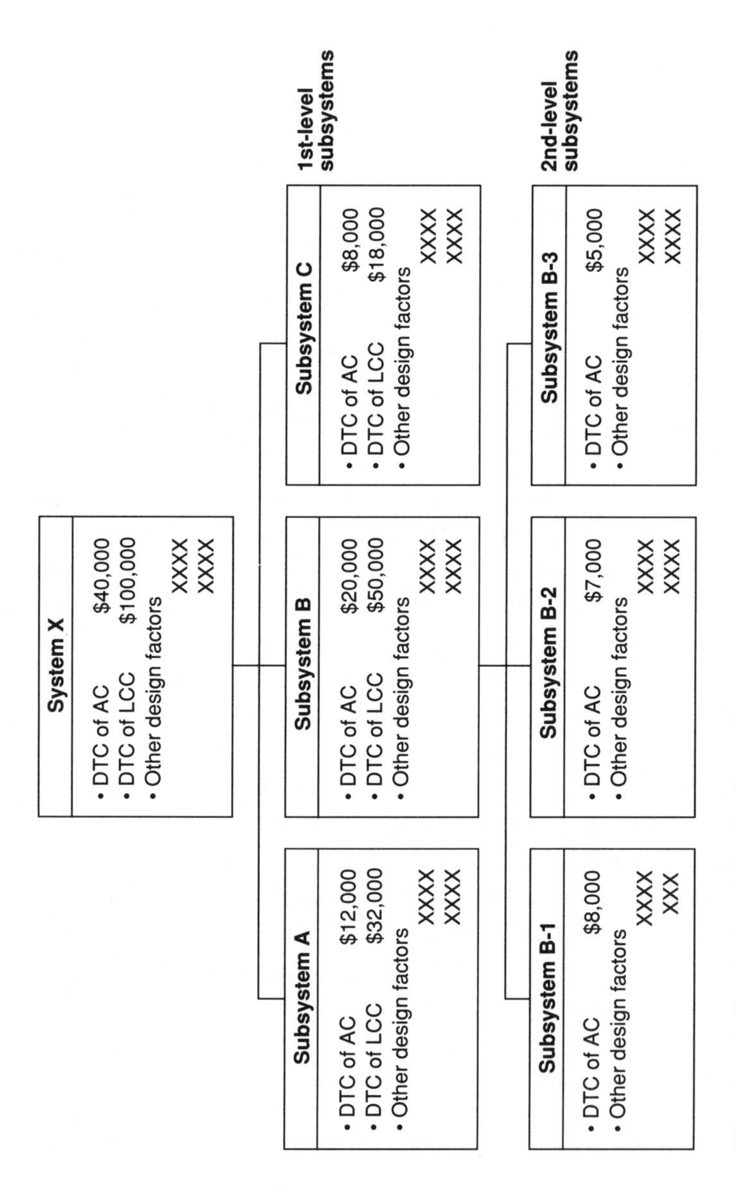

Figure 4-1. Design-to-Cost (DTC) Outline

costs among the first-level subsystems, then among the second and subsequent levels, until they are spread over all subsystems. Figure 4-1 outlines this cost distribution process.

The total cost, however, is often difficult to comprehend. Figure 4-2 shows that only the tip of the cost iceberg is visible. The initial (acquisition) costs are easy to see, but the running costs are not. Failure to consider running costs can lead to many problems. At least 80 percent of a system or equipment unit's LCC is determined at the conceptual design and preliminary system design stages (Figure 4-3), which is why LCC design parameters should be worked out at the design stages.

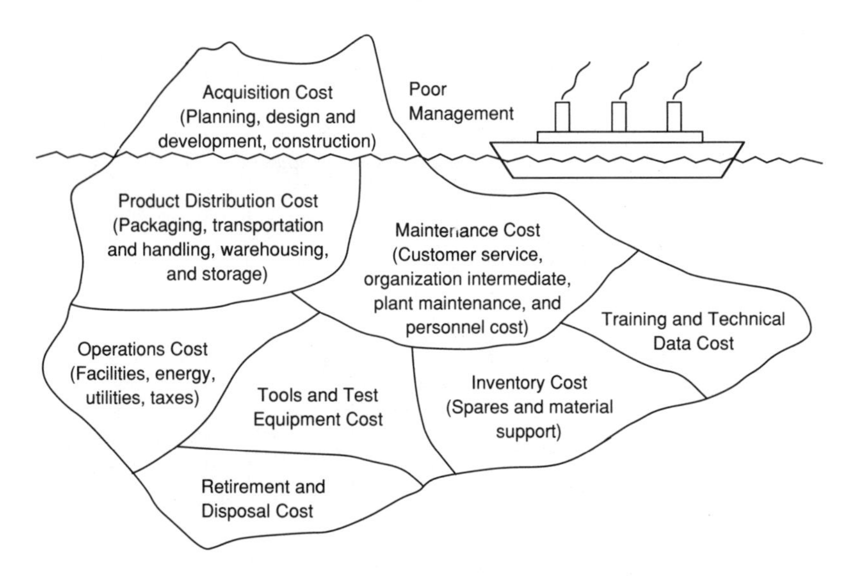

Figure 4-2. Total Cost Visibility

BASIC APPROACH TO LCC DESIGN

Ideally, equipment should be maintenance and problem-free with minimal life cycle costs, but these desirable features are mutually exclusive in some ways. Therefore, we must find

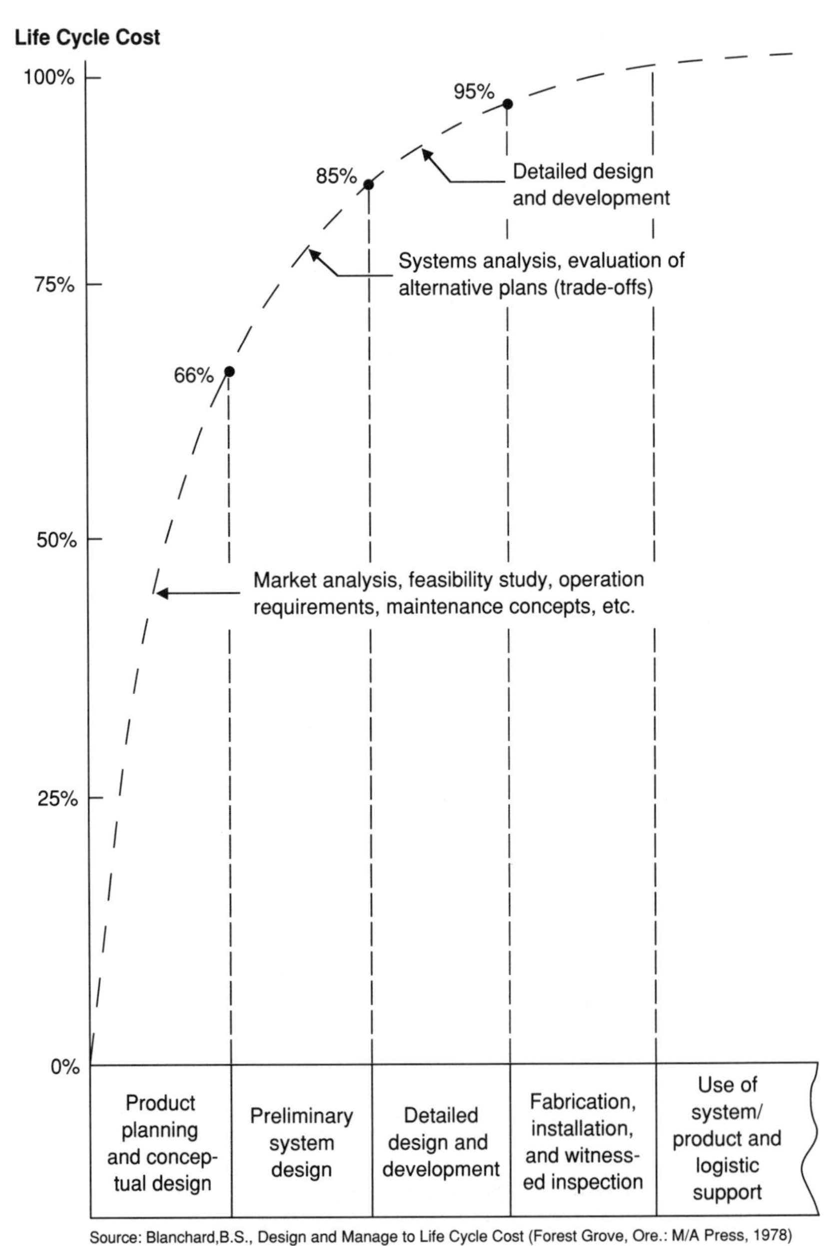

Source: Blanchard,B.S., Design and Manage to Life Cycle Cost (Forest Grove, Ore.: M/A Press, 1978)

Figure 4-3. Business Factors Influencing Life Cycle Cost

an LCC design that provides the most economical trade-off between them.

When working out the LCC design, answers to such basic questions as what equipment to design and what kind of design to create depend largely on the design mission (product quality and cost targets, technical issues, and so on).

Accurately calculating the life cycle cost is not the purpose of LCC design. It is rather an optimal design approach in which life cycle cost is just one of several design parameters. Estimating the LCC is a method for discovering and elucidating the requirements of the best possible design via the shortest possible route. Accordingly, in considering the design questions posed by the design mission, first determine the goals for the equipment and the acceptable range for LCC.

For instance, suppose the equipment to be designed includes a multi-welder, which directly processes product, and a boiler, which supplies energy to the production equipment. Clearly, these two pieces of equipment serve different purposes. The multi-welder supports production capacity and product quality, primarily, while the boiler provides energy efficiently in support of the energy supply capacity.

The multi-welder's LCC lifetime range (number of years of use) can be determined by the product life cycle (the production period). Other criteria determine the boiler's useful life, however, such as the expected economic life or the period before the boiler is rendered obsolete by technological advances. So, the first step is to establish an LCC design outline aimed at overcoming technical obstacles to achieving the equipment goals and LCC range based on the design mission.

You must answer the following questions when implementing this approach:

- What costs constitute the LCC?
- What LCC estimation method is the most appropriate, and when in the design process should we perform LCC estimations?

- How can we evaluate the appropriateness of the LCC?
- What procedure shall we follow in working out the LCC design?
- What data and standards are needed for LCC estimations and LCC design?
- How do we work into the evaluation intangible factors not included in the LCC estimation?

General Procedure for LCC Design

The basic elements of LCC design are the design-to-cost approach and the various units or subsystems among which costs are distributed (parts of production lines, certain processes or equipment units, and so on).

For now, use the following simple equation:

$$\text{LCC} = \text{IC (initial costs)} + \text{RC (running costs)}$$

while noting, however, that the IC and RC vary considerably depending on the design mission.

Figure 4-4 outlines the following six steps in the basic LCC design process.

- Step 1: Conduct initial cost reduction (ICR) design to minimize initial costs.
- Step 2: Calculate LCC.
- Step 3: Clarify running cost reduction (RCR) needs by examining the highest running cost items.
- Step 4: For RCR design, draft improved (alternate) plans that also minimize running costs.
- Step 5: Estimate additional IC and RCR effects in each alternate plan.
- Step 6: Trade off; apply engineering economics principles to select the alternate plan that offers the best DTC or calculated interest.

Case 1: Minimum IC design

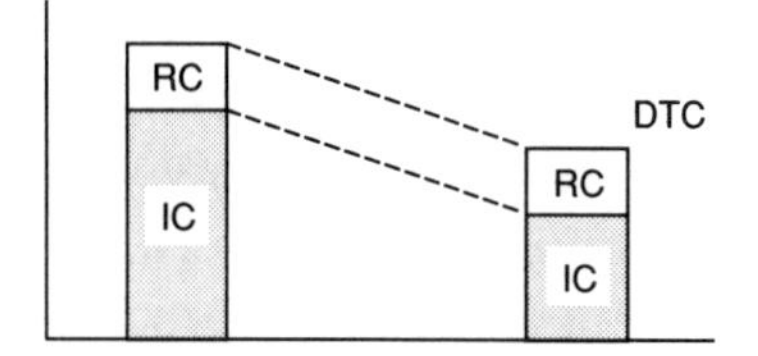

Case 2: Minimum RC design

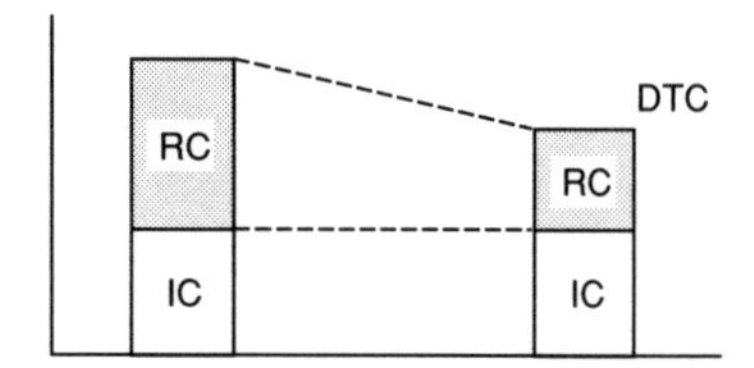

Case 3: ICR-RCR design

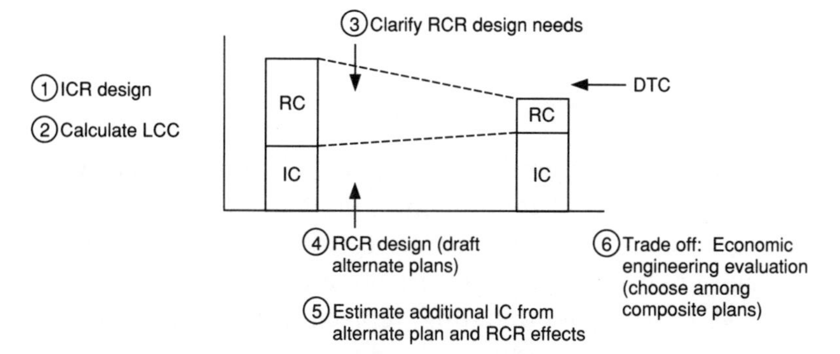

Figure 4-4. Six Steps in the Basic LCC Design Process

The goals and restrictions that management places on LCC design, the required equipment characteristics, and the established design mission will determine when each of the following three LCC design approaches applies.

- Case 1: Attempts to minimize IC without raising RC
- Case 2: Attempts to minimize RC without raising IC
- Case 3: Attempts to meet targets for both IC and RC (i.e., minimum LCC design — the basic goal of LCC design)

MINIMUM IC DESIGN

No designer can hope to escape the principles of economics. In design you must spend money to make money, but good engineers spend carefully, as if the cost were coming out of their own pockets. In a good design you do not strive for perfection but to provide all the required functions at the lowest cost. Most designers must deal with various restrictions, however. For example, new equipment designs cannot completely escape from the conceptual framework of existing equipment, production systems, and manufacturing methods. Moreover, designers naturally work for interesting, novel, and refined designs, but these factors often add cost.

Functional Cost Needs Analysis

Functional cost needs analysis (FCNA) design is a type of LCC design aimed at minimizing initial costs (without raising running costs) within the existing conceptual framework. This approach analyzes equipment function and seeks out the least expensive means of eliminating, replacing, or fulfilling detailed functions. The basic method behind the FCNA design approach is called the function/needs (FN) method.

In the FN method, *function* refers to the equipment function (application) that serves the main system or product under development. *Needs* refers to the various needs or conditions that must be met by that system. By distributing responses to these needs among the various functions using a matrix charting method (see Figure 4-5), you can work out systematically the characteristics and specifications that will improve the required functions.

Basically, the FN method involves four activities:

1. Defining system functions (F)
2. Defining system needs (N)
3. Distributing needs among functions and determining specifications
4. Determining the combination and specification of conditions required from the system needs

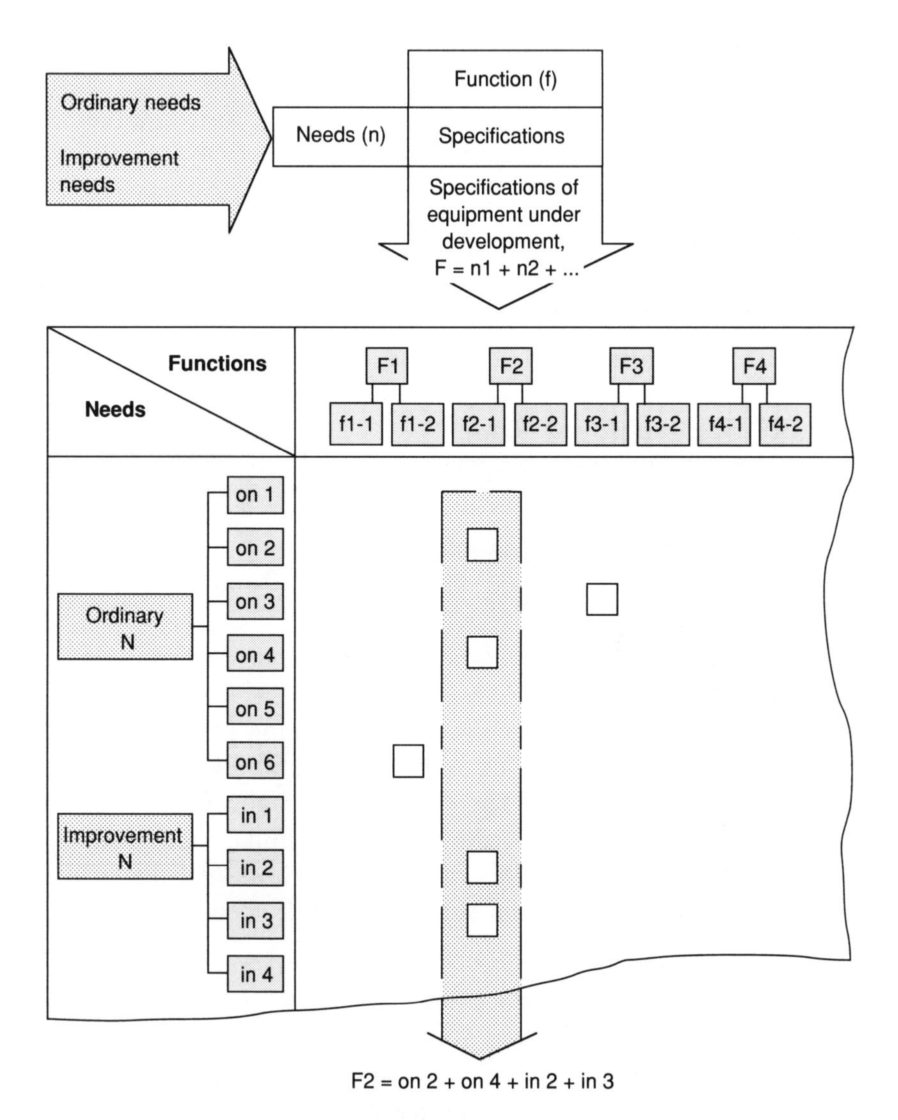

Source: Yoshio Ono, *Planning to Introduce Robots* (in Japanese) (Tokyo: Nikkan Kogyo Shimbun, 1988).

Figure 4-5. Outline of F-N Analysis Method

Case Study 4-1: FCNA Design Procedure at Toyota Auto Body

The following case describes how the FCNA design procedure was used by a team at Toyota Auto Body in a press die design process. The FCNA procedure began with the selection of design targets and ended with the establishment of an optimum plan. As such, the procedure fell into the following four stages (see I-IV, Figure 4-6):

I. Create function cost analysis table.
II. Discover improvement needs.
III. Draft improved plans.
IV. Evaluate improved plans.

I. Create Function × Cost Analysis Table

There are two steps in the first stage.

1. Develop equipment functions. A die design may serve several functions. It can add value to products; it can also contribute to the production function by enhancing productivity, to the maintenance function by increasing durability, shortening repair times, and so on. The team studied the press die for which it intended to lower costs along with similar dies currently in use. The team listed the various functions of these dies, breaking the functions down as far as the unit-specific level.

2. Analyze function-specific costs. The team analyzed how function costs that had been broken down into unit-specific function costs were involved at each stage of design and development, from conceptual design to design drafting, fabrication, test run, process processing, and maintenance.

At this point, the team created a function × cost analysis matrix (shown in Figure 4-7) and then calculated the function × stage-specific costs based on the labor hour reports,

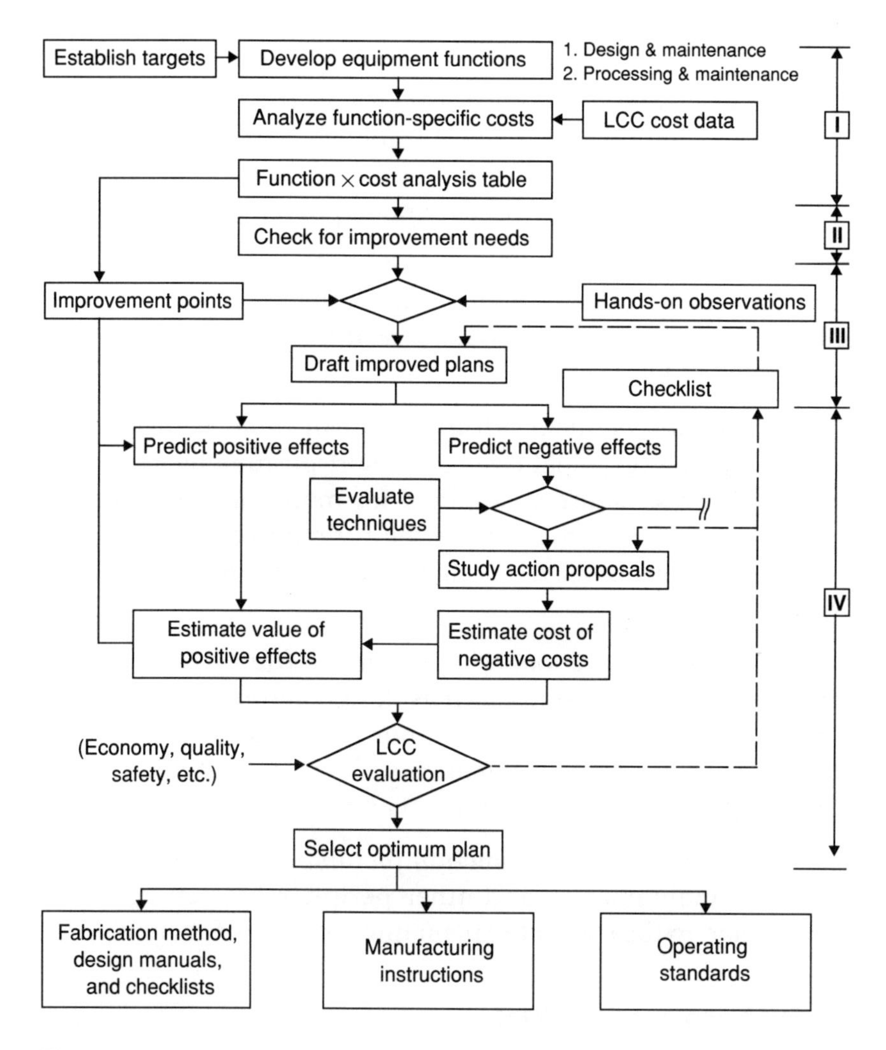

Figure 4-6. Procedure for Function × Cost Analysis

procurement prices, daily operating log, maintenance records, and other relevant data. The team filled in these cost figures under the appropriate factor column in the matrix.

II. Discover Improvement Needs

After creating the function × cost matrix, the team began rooting out improvement needs (problem points) based on the matrix. It discovered that the following four points deserve emphasis with regard to discovering improvement needs.

1. ***Appraisal of functions per se.*** Ascertain whether the function really helps in terms of quality, utility, and maintainability. Is the function necessary, or is it retained simply because it has always been there?

2. ***Balance between initial costs and running costs.*** If the design is long on initial costs and short on running costs, perhaps the die structure is designed for more functions than are really necessary. Conversely, if the running costs clearly outweigh the initial costs, the design may not ensure sufficient reliability.

3. ***Cost balance among steps in the life cycle.*** Are there imbalances between steps such as design and fabrication, material costs and manufacturing labor hours, or manufacturing products and making adjustments? Working out such imbalances in the design can lower the manufacturing cost.

4. ***Balance between function factors.*** Are the costs of production, maintenance, and other peripheral functions out of proportion to the cost of the main unit's primary function?

III. Draft Improvement Plans

After the team identified the improvement needs, in-house specialists in design, fabrication, and maintenance met to review the list of improvement points, study drawings, and brainstorm improvement ideas. Table 4-1 lists points to consider in improving a drawing press die. In drafting this list, the participants used brainstorming checklists and various industrial engineering principles to generate ideas for common equipment items.

Table 4-2. Drawing Press: Disadvantage Checklist

Reliability	Can the between-maintenance periods be shortened? Do the dies loose or wear easily? Have the maintenance methods (parts, frequency, procedures, causes) been clarified? Can press operators or maintenance staff predict abnormalities easily?
Product Quality	Does quality ever vary widely? Is the forming process ever unstable? Are products damaged during handling? Are products scratched? Is the punching process ever inadequate? Do products crease, crack, or warp easily?
Maintainability	Does the equipment break down easily when operator errors occur? Do breakdowns often result in major problems? Can breakdowns be restored quickly? • Is the equipment easy to take apart and reassemble? • Are the parts and materials interchangeable? • Is it easy to find the causes of breakdowns? • Does the maintenance staff have the proper skills to respond to problems? Can breakdowns be restored cheaply (can spare materials and parts be purchased cheaply)? Do maintenance staff attempt to predict abnormalities? Is the equipment in line with all regulations and in-house standards?
Safety, Hygiene, Environment	Does the work involve major muscle strain (e.g., in feeding and removing workpieces)? Is the work posture unhealthy? Do breakdowns cause major cost problems? Is the noise sometimes beyond a tolerable level? Are dust and/or foul odors ever a problem? Are any special protective measures required? Does disposal of old equipment require any special procedures?

improvement plan might incur in terms of quality and maintainability. They found that under certain conditions (for example, if the number of units were left as is) the improvement plan could be adopted without incurring too many problems later.

Example 2: Pull System for Material Supply

Cost factors such as breakdown loss and maintenance are closely and clearly linked to the die's operability functions. Although such costs arise when these functions are unreliable, in this case the most effective improvement lay in replacing the functions with general purpose functions by switching from the push method of materials supply (P) to the pull or vacuum method (V). Since this improvement plan required the remodeling of general purpose equipment, the designers conducted an LCC evaluation of the plan with an emphasis on quality and safety and found a cost advantage in the final analysis, which prompted them to adopt the plan.

Advantages of FCNA Approach

This example illustrates that the best approach lies in comparing functions to corresponding costs. In other words, by examining the functions closely and working out their costs using function × cost analysis matrices you can uncover hidden waste. In the team's improvement planning it went beyond the die itself to study possible improvements in the fabrication methods, production system, and other areas. A thorough evaluation of such comprehensive improvements helped the team anticipate and overcome disadvantages that otherwise would have been overlooked. Another advantage of this approach is that it helps clarify technical issues that deserve future study. Addressing and resolving these issues raises the level of your technical knowledge and can easily lead to major cost reductions.

Figure 4-8 shows the results of the improvement activities described above. Both the weight and area of the die were substantially reduced, resulting in a 28 percent savings in fabrication costs. This by itself was almost enough to achieve the cost reduction target for the new product. Figure 4-9 compares the weights of the conventional die and the new die. More attention to simplification and process conservation led to a more compact (waste-free) die structure. As mentioned above, the result was a major reduction in the die's weight and surface area.

MINIMUM RC DESIGN

Maintenance prevention design aims to maximize reliability, maintainability, operability, and safety. Thus, it can be used as an LCC design method to realize greater savings by minimizing running costs (without raising initial costs). The following discussion focuses on how minimum RC design can be applied as an LCC design method to reduce breakdowns.

In this context, minimum RC design involves clarifying the breakdown characteristics, finding out which breakdown-preventing measure works best, and then working out a design while verifying the measure's economy (including its positive or neutral impact on initial costs).

A time-series analysis of breakdown events reveals a bathtub curve, as shown in Figure 4-10. In this case, the goal is to achieve a minimum RC design that reduces early breakdowns and routine production breakdowns (those that occur during regular production).

In addition, the following perspective is helpful in designing to reduce wear-related breakdowns:

- To reduce or limit initial costs, extend the life of the equipment's parts to approximate the life cycle of the equipment.

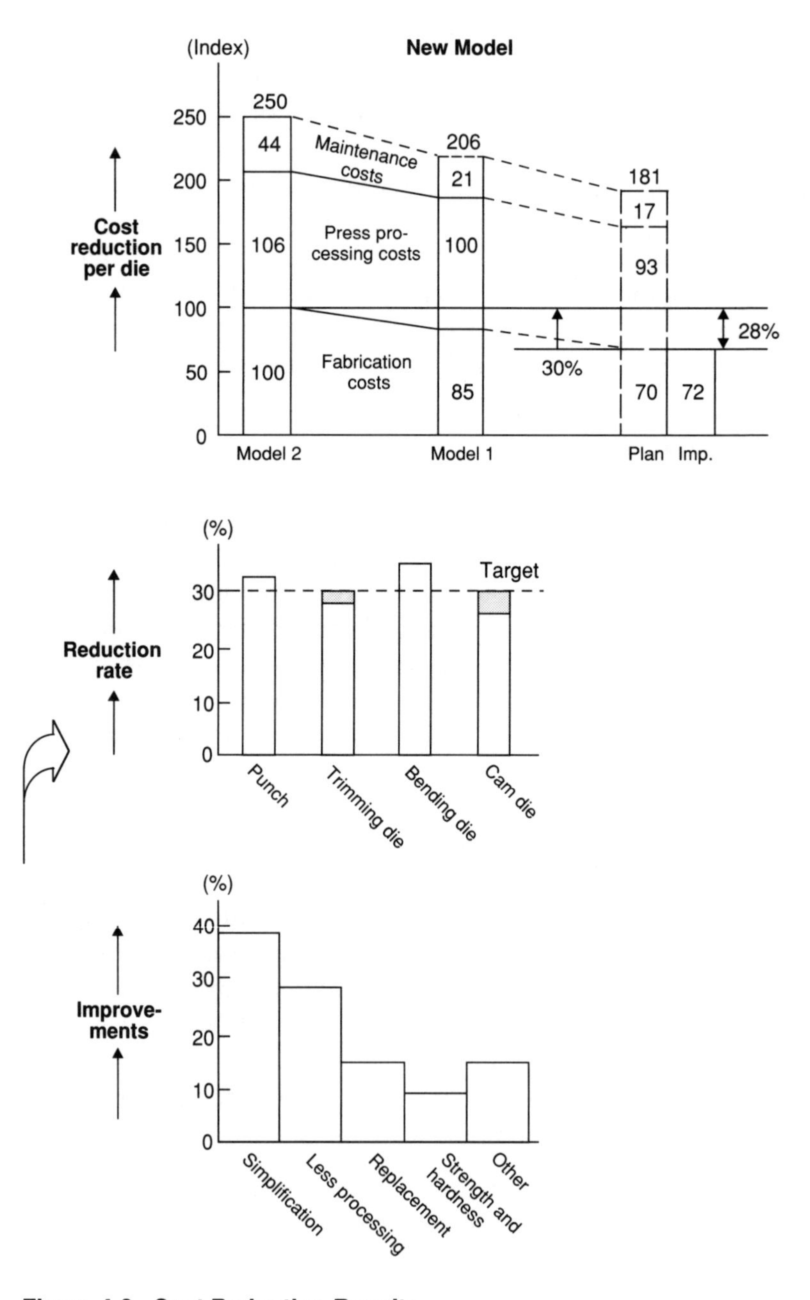

Figure 4-8. Cost Reduction Results

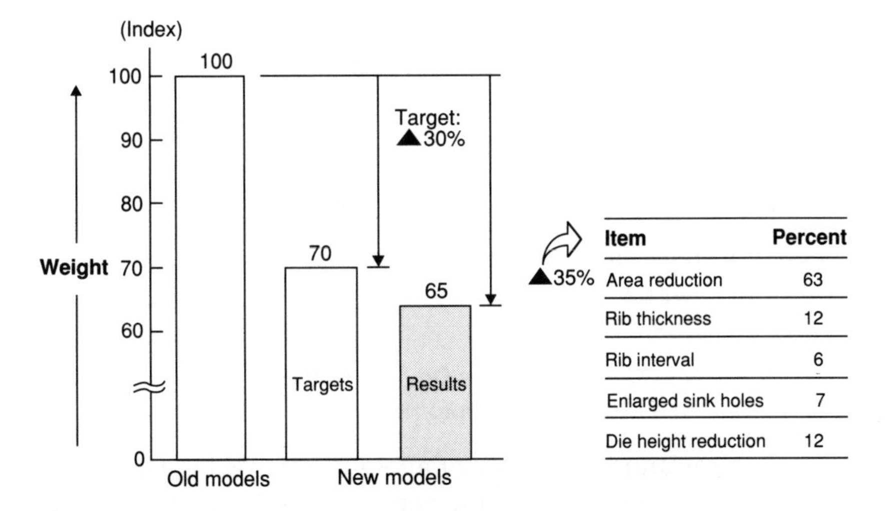

Figure 4-9. Die Weight Reductions (B Class or Above)

- Address the problem of wear-related breakdowns by reducing breakdowns that occur during regular production.

Is an inexpensive design that produces equipment with a shorter life better than an extremely costly design that ensures a longer life? Generally, keeping a large stock of spare parts and risking the loss-related costs of unexpected major breakdowns is hardly conducive to minimum RC design. To achieve minimum running costs without raising initial costs may require a design

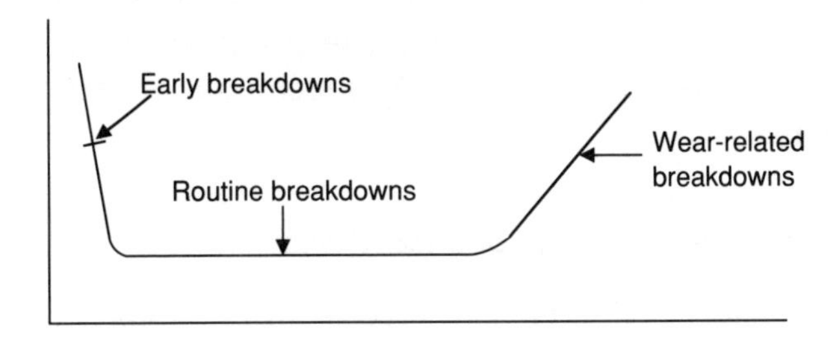

Figure 4-10. Breakdown Events

with a shorter life cycle that nevertheless minimizes costs by enabling easy prediction of periodic parts replacements.

Early Breakdown Characteristics and MP Design

When new equipment starts breaking down, the problem generally stems from one of the following five situations (introduced in Chapter 3).

1. New Product Characteristics Do Not Match Fabrication Methods and Equipment

If, for example, you cannot avoid using existing fabrication methods and equipment for the new product, the design is likely to present some problems in terms of precision and strength. In such cases, begin during the early stages of product design by analyzing the early breakdowns of existing equipment and understanding the abnormalities behind them. In other words, begin with the goal of creating a product design free of breakdown-producing factors. Consider preventive measures before drafting the product design drawings since it is harder to make design changes once the drawings are complete.

2. Problems with New Fabrication Methods or Equipment

By contrast, using new fabrication methods or new equipment can also be a source of trouble. This type of problem can be prevented by conducting simple preliminary tests or by using computer-aided design (CAD) techniques to simulate the equipment.

3. Errors in Detailed Design of Small Parts

Keep lists of "don'ts" garnered from previous early breakdowns, designer mistakes, and so on. Incorporate them in checklists so designers can avoid repeating mistakes. These lists can serve as the basis for a thorough design check at the preliminary evaluation stage.

4. Errors during Fabrication, Installation, and Startup Stages

You can use several methods to prevent errors during the fabrication, installation, and start-up stages: preliminary evaluation for the witnessed inspection stage, checklists for careful review, and CAD/CAM to help ensure error-free fabrication. Many problems that arise during the installation and startup stages are related to items like limit switch setup and wiring layout, which are hard to represent on the design drawings.

It is risky to leave everything up to people conducting the startup stage. Instead, standardize installation and startup procedures and support them with hands-on training and meticulous checking. Even at the test run stage, you should be checking the equipment's weak spots and other problems identified by the preliminary evaluation and conducting overload tests.

5. Errors in Operating and Maintenance Procedures

Enlist the participation of the operations and maintenance staff early, at the equipment design stage, and use their input to better understand the equipment's functional structure. Begin training the operations and maintenance staff in the correct procedures as early as possible.

The following examples illustrate how MP design can be developed to reduce early breakdowns.

Example 1: Working from the Product Design Stage

Traditionally, the design review is conducted once the structural drawings are finished, and it reveals the need for product modifications to accommodate the equipment or method of operations. Structural modifications are difficult to make at this point, however, and in many cases the modification proposals are not adopted.

To address this problem, organize and consider at the conceptual planning stage all early breakdown data and MP data for existing products (and corresponding production equipment) that relate to the new product design. This generates a ready-made preproduction checklist. This preproduction check can eliminate causes of breakdowns by reducing the amount of required equipment through consolidation or simplification of products, and by simplifying the equipment structure. Figure 4-11 shows how a preproduction checklist can be applied in improving a product design.

Example 2: Improvements at Equipment Design and Fabrication Stages

This example focuses on the review and improvement of installation standards for wiring and piping — items that are especially prone to problems at the initial operation stage. All the required wiring and piping conditions are clearly organized in Figure 4-12 to show how abnormalities have manifested themselves in each functional part of the equipment. This analysis clarifies how breakdowns have occurred in existing products (and corresponding production equipment) under specific terminal connection conditions. Furthermore, it uses a matrix format to minimize the risk of omissions by showing how the various hoses and leads should be installed.

Using this diagram, the designers generated improvement proposals that thoroughly prevent even minor defects in parts where only one abnormality had occurred. This method clarifies what steps must be taken for each part of the equipment to meet the required conditions; it provides data that are easier to understand when training equipment fabrication workers to install the wiring and piping; and it results in wiring and piping

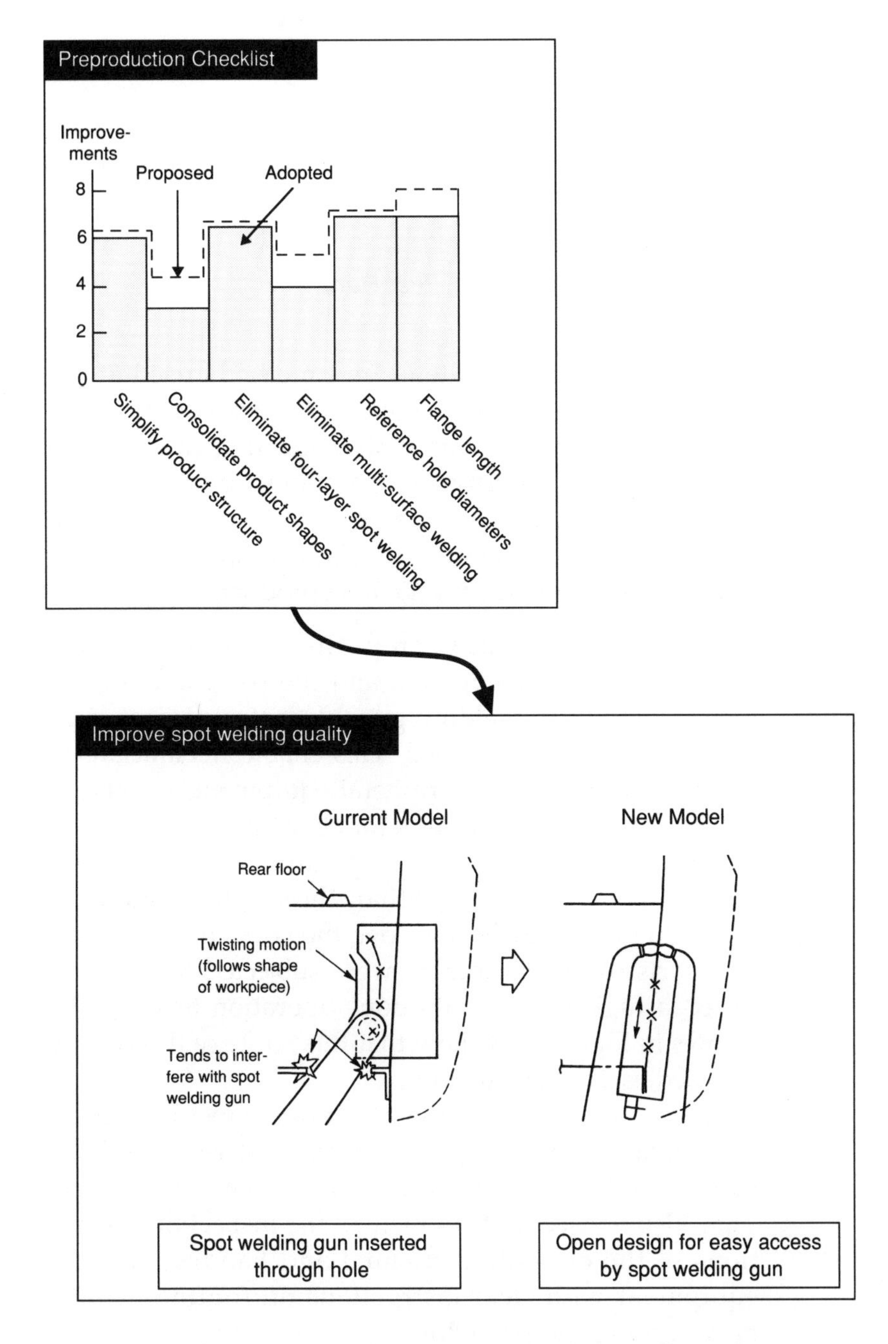

Figure 4-11. Preproduction Checklist and Application Example

that not only meet all functional requirements but are easier to check and repair.

Example 3: Improved Installation, Test Run, and Handover Methods

Typically, the evaluation items and methods used at the installation, test run, and handover stages are inadequate. Thus, a lot of time must be spent hurriedly collecting data, which increases opportunities for errors in operations manuals. This in turn, increases the time it takes to respond to abnormalities that emerge at the early operation stage. In this example, the following improvements were made for the new product:

a. At the final fabrication step, the connection status between the base equipment and the peripheral equipment (conveyors, welding guns, transfer devices, etc.) has already been confirmed. This enables technicians to build dummies of the peripheral equipment, which can be simulation-tested to check for abnormalities.
b. During previous continuous operation tests, time constraints resulted in a less-than-thorough exposure of early abnormalities. To improve the correction rate, certain parts of the equipment were subjected to overload or extra high-speed continuous operation testing. The arrows in Figure 4-13 show that this strategy thoroughly exposed early abnormalities.
c. Based on the results of the continuous operation tests, technicians made improvements that rendered the equipment less likely to break down and easier to repair. They also revised and augmented the inspection instructions in the operations manual and handed over the equipment with clear instructions on routine maintenance timing and procedures.

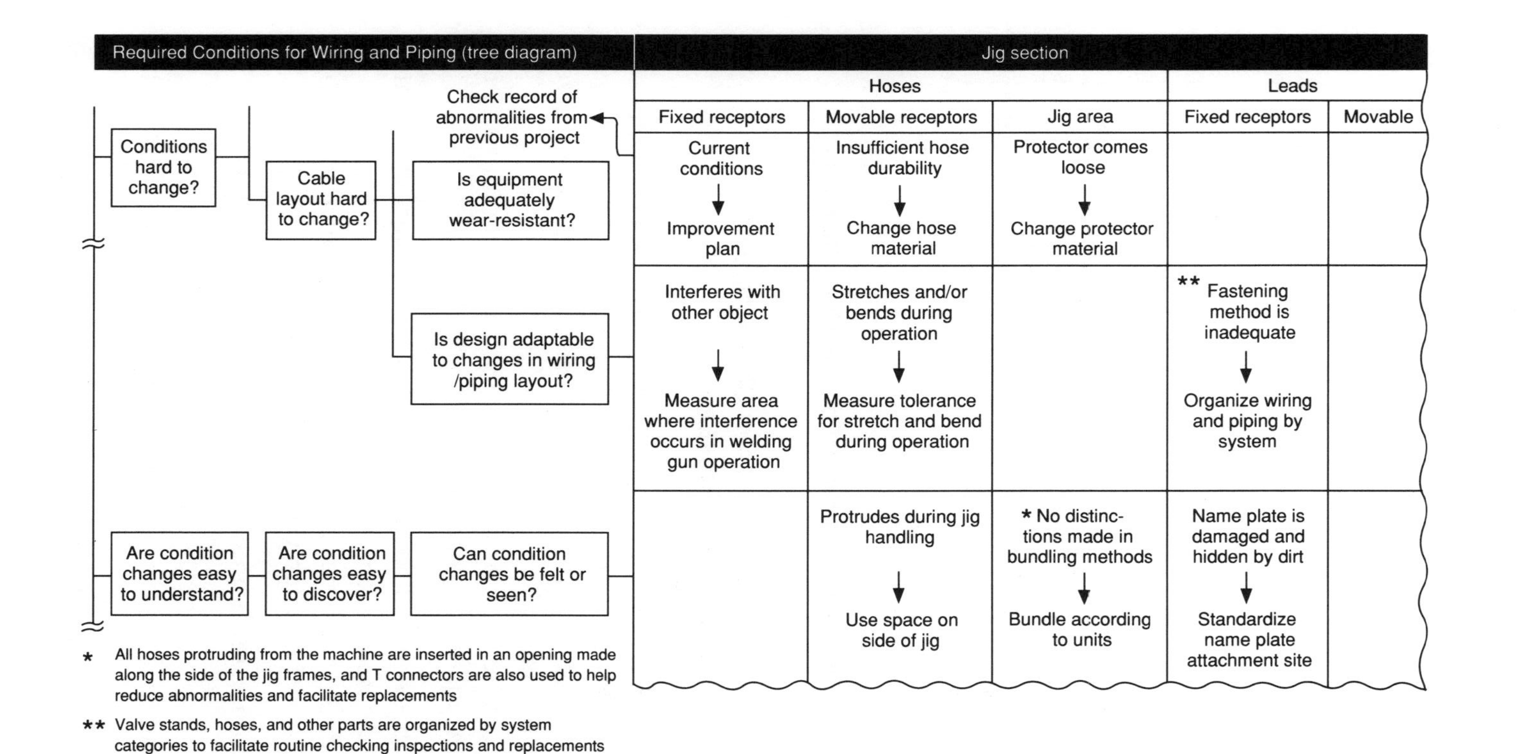

Figure 4-12. Organization of Wiring and Piping Conditions

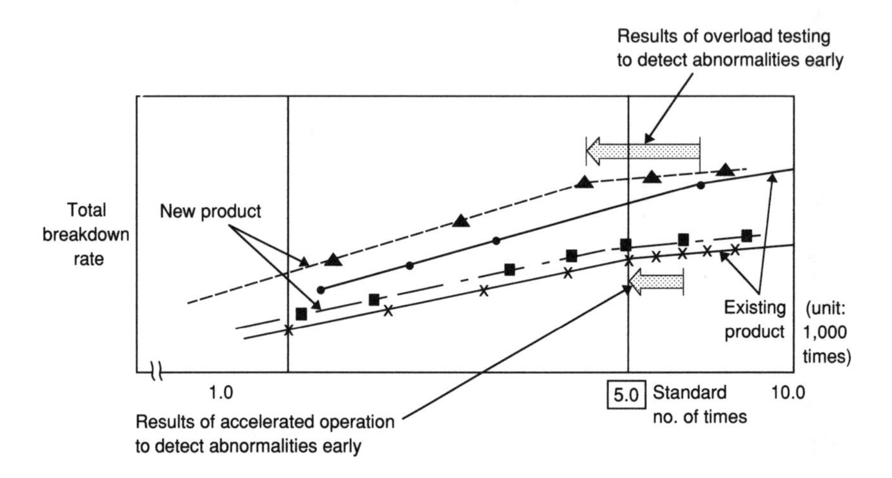

Figure 4-13. Review of Continuous Operation Test Standards

Routine Production Breakdown Characteristics and MP Design Countermeasures

Once production has stabilized, breakdown characteristics can be divided into two broad categories: *sporadic breakdowns* that occur randomly and *chronic breakdowns* that occur at relatively short intervals.

Sporadic breakdowns generally arise from human error under the following conditions:

- Inadequate knowledge or skill on the part of operators and maintenance staff
- Poor equipment operability or maintainability, which almost inevitably leads to human error

Chronic breakdowns are the result of the following situations:

Inadequate prediction of part life. Since the life of certain parts turns out to be shorter than predicted in the design (designs often ignore use conditions for general-purpose parts), the maintenance schedule is too slow, and this leads to chronic breakdowns.

Unclear inspection and repair procedures. Even when the life of the parts has been accurately predicted, chronic breakdowns occur when routine checking and repair procedures are not spelled out clearly.

Poor maintainability. The equipment's poor maintainability makes routine inspections difficult, if not impossible, and makes deterioration hard to spot.

Load imbalances. Load imbalances not foreseen at the design stage (such as variance in processing conditions or workpiece characteristics) result in shorter part lives. In such cases, simply replacing parts is not enough to prevent chronic breakdowns.

You can use MP data effectively to prevent production breakdowns — if you take the following steps:

1. Predict breakdowns at the design stage.
2. Deduce the causes of predicted breakdowns.
3. Change the design to eliminate those causes. Prepare alternate plans to address the deduced causes from the standpoint of reliability, maintainability, and operability.
4. Next, select from among the alternate plans, basing the decision on cost.
5. After choosing, decide which steps should be taken at which design/development stages (step-by-step management).
6. Finally, clearly describe the required maintenance methods in the operations and maintenance manuals.

Now take a closer look at the first three steps.

Predicting breakdowns. Predictions can be based on breakdown records for similar equipment or parts, or they can be based on fault tree analysis (FTA) or other prediction methods.

(This latter approach is best when designing fundamentally new equipment or parts.)

Deducing breakdown causes. Causes can be deduced using the "five whys" approach (asking "why" at least five times to get at the root cause), or through P-M analysis.*

Proposing alternate plans. Before proposing an alternate plan, it is essential to investigate the plan's reliability and maintainability. To ensure that the plan will be effective in both these areas and to facilitate a balance between them, evaluate the plan in terms of the following five conditions for stable operation. In this evaluation, the five conditions for stable operation are simply a breakdown-preventive rephrasing of the five conditions for QA-friendly equipment described in Chapter 3. The first three conditions are related to reliability.

- Are conditions for preventing breakdowns clear?
- Are conditions easy to establish?
- Are conditions resistant to change?

The remaining two conditions are related to maintainability.

- Are changes in the conditions for preventing breakdowns easy to recognize?
- Are conditions easy to restore?

The following case studies illustrate these approaches in detail.

---

* In P-M analysis, phenomena associated with a failure or defect are thoroughly analyzed in terms of the actual physical principles behind them. Every factor involved is then studied in detail. Factors found to be off-specification are restored, and control specifications are established and maintained for factors lacking them. For a more detailed explanation of this improvement method see *Training for TPM: A Manufacturing Success Story* (Cambridge: Productivity Press, 1990).

Case Study 4-2: MP Design Countermeasures for Routine-production Breakdowns

Figure 4-14 shows PM data from an analysis of chronic stoppages in equipment used to produce an earlier product during an 18-month period in a fixed location. As shown in the figure, various breakdown phenomena appear to have occurred randomly, but most were due to recurring problems. Even novel breakdown modes, however, were analyzed as being fully solvable under existing technology. This analysis led the engineers to incorporate full-scale preventive measures in the design of the new product (and corresponding production equipment).

Because most breakdowns occurring in this fixed-location equipment were short stoppages that recurred frequently, the engineers set a goal of achieving a MTBF rate for the new product and equipment that was five times greater than the previous rate.

Implementation of FTA and P-M analysis

In the past, this company found that using fault tree analysis (FTA) in the preliminary evaluation of equipment reliability produced some positive results. When they added P-M analysis of new products, with reliability and maintainability analyses based on the five conditions for stable operation, their preventive measures were free of oversights in evaluation. Figure 4-15 illustrates schematically the product development process for this case.

Section *(a)* of the figure shows how the conventional FTA approach was used to find the breakdown modes down to the parts level. Section *(b)* analyzes an earlier product's breakdown estimates and results. Section *(c)* is a P-M analysis of breakdown modes manifesting a large gap between breakdown estimates and results and newly discovered breakdowns.

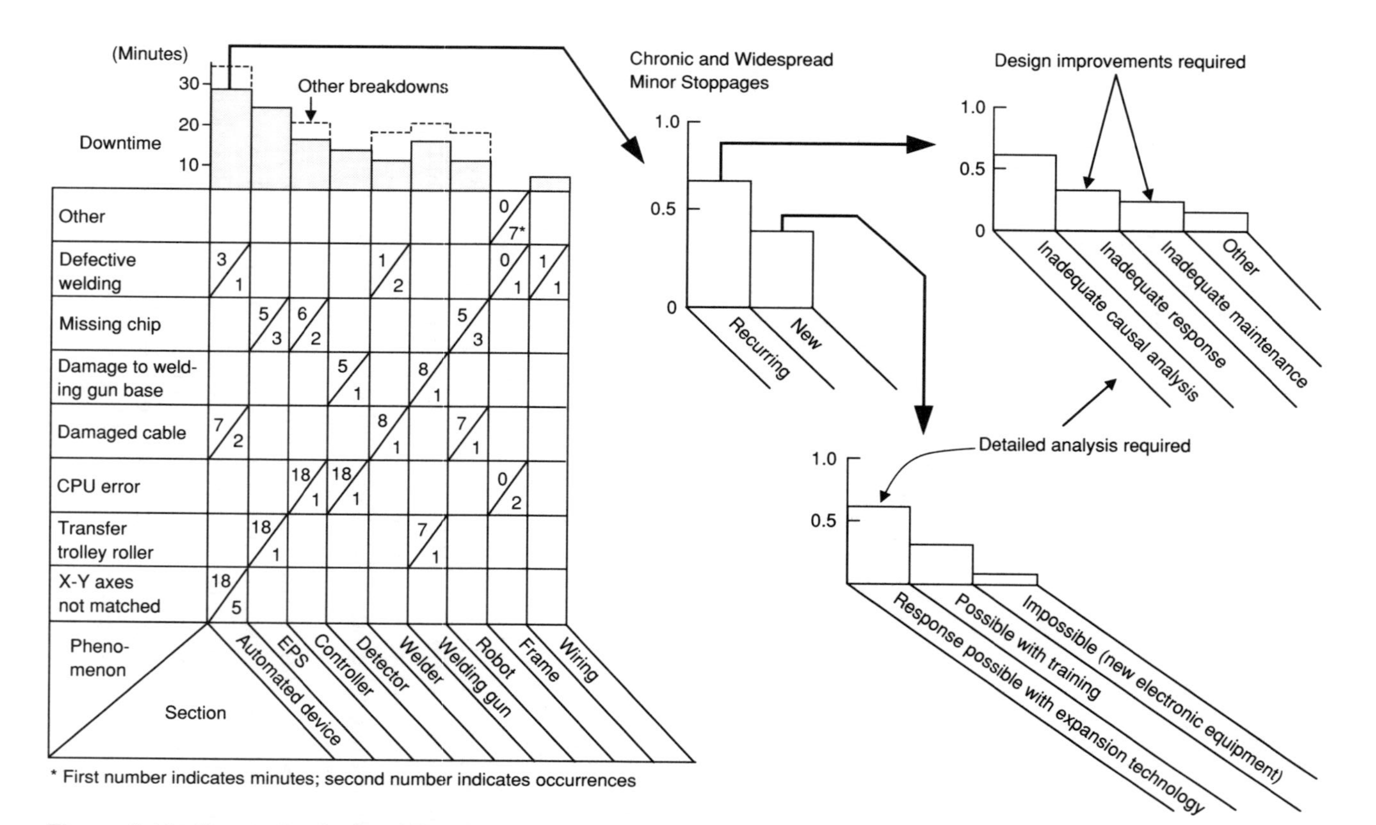

* First number indicates minutes; second number indicates occurrences

Figure 4-14. Cause Analysis of Routine Production Breakdowns

The engineers drafted several plans to effectively treat all the physical causes of these breakdown modes. Then they evaluated the countermeasures using the five conditions for stable operation. Their evaluation based on the three reliability conditions was recorded in section *(d)*, and their evaluation based on the two maintainability conditions was recorded in section *(e)*. Next, the engineers refined their countermeasures and developed more detailed improvement plans. Section *(f)* shows the results of economic evaluation and the organization of implementation steps, as well as the results of a study confirming reduction predictions and the plan's feasibility.

Using this FTA/P-M analysis system in its improvement activities, the company was able to plan, execute, and realize benefits from various improvements aimed at many types of minor stoppages. In addition, their five-condition evaluations led to increased safety, which was related to the reduced impact of random defects and higher reliability in welding methods.

Figure 4-16 illustrates an improvement generated through this system involving a small automatic spot welder — the device with the highest breakdown rate during production of an earlier product. The production equipment used for the previous model had a complicated structure with many components, which made it prone to breakdowns and difficult to maintain. By contrast, production equipment for the new model benefited from P-M analysis and evaluation based on the five conditions for stable operation. The company radically revised the production equipment's structure and developed a much simpler biaxial automatic welder to do the job for the new product.

The main improvements included changing the drive system (a major source of breakdowns in the previous equipment), simplifying the structure of certain mechanisms to boost reliability, selecting more reliable parts, eliminating certain limit switches (originally intended to prevent deterioration of maintainability) that had caused intermittent

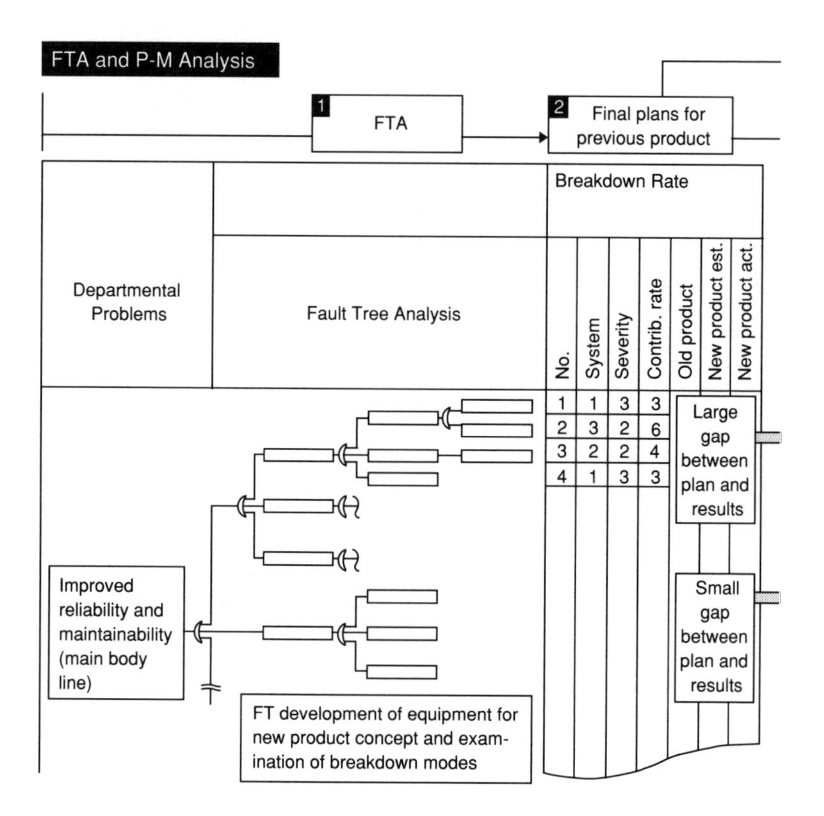

Figure 4-15. Procedures for Fault Tree Analysis (FTA) and P-M Analysis

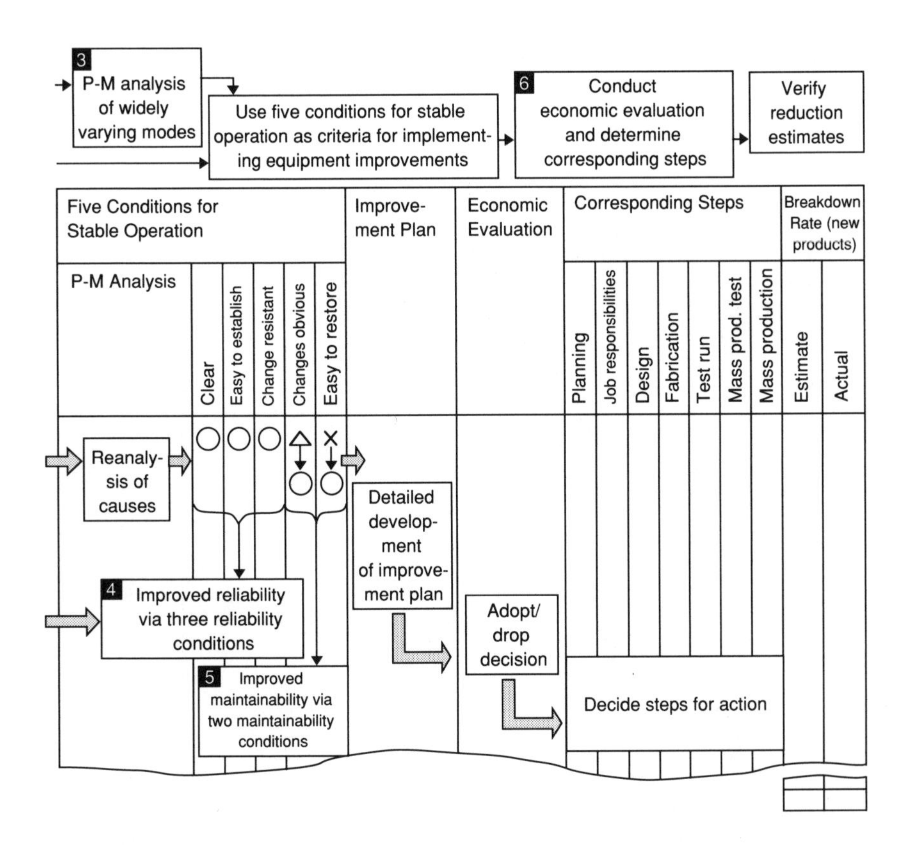
3
P-M analysis of widely varying modes
Use five conditions for stable operation as criteria for implementing equipment improvements
6
Conduct economic evaluation and determine corresponding steps
Verify reduction estimates
Five Conditions for Stable Operation
Improvement Plan
Economic Evaluation
Corresponding Steps
Breakdown Rate (new products)
P-M Analysis
Clear
Easy to establish
Change resistant
Changes obvious
Easy to restore
Planning
Job responsibilities
Design
Fabrication
Test run
Mass prod. test
Mass production
Estimate
Actual
Reanalysis of causes
Detailed development of improvement plan
4
Improved reliability via three reliability conditions
5
Improved maintainability via two maintainability conditions
Adopt/ drop decision
Decide steps for action

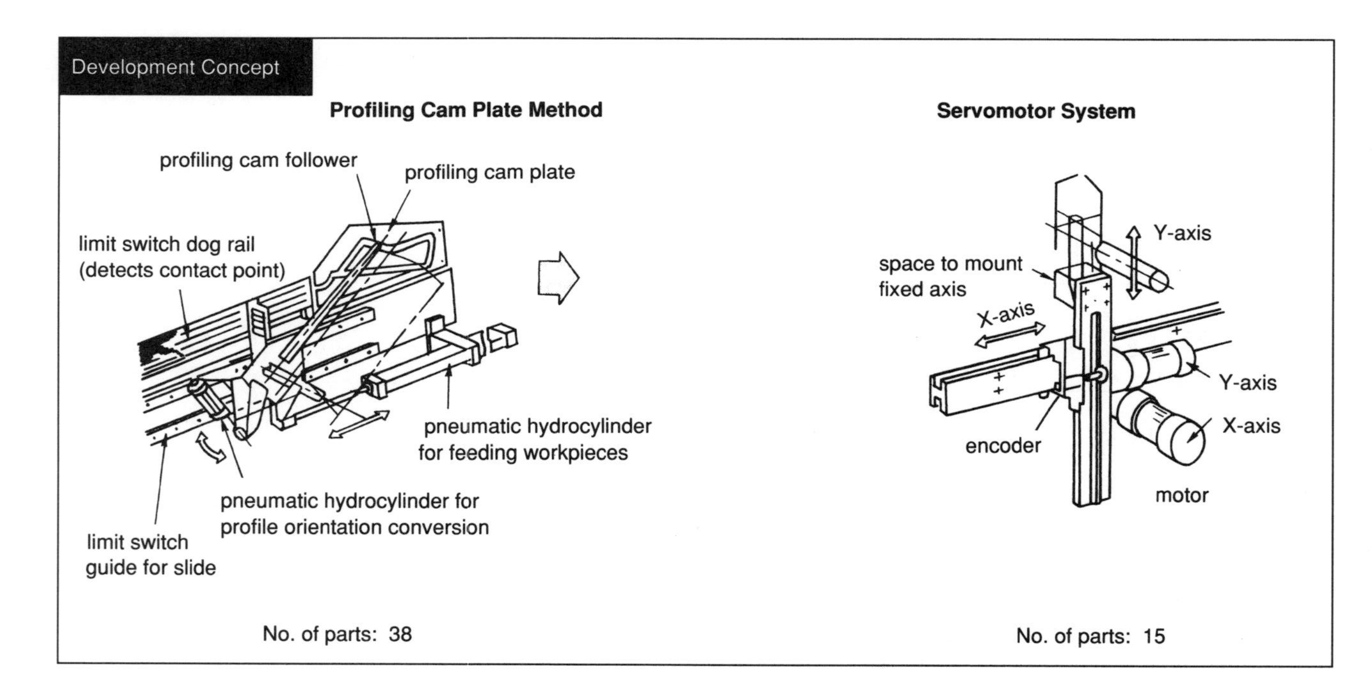

Figure 4-16. Development of Small Biaxial Automated Device

stoppages, and halving the number of components. These and other improvements collectively reduced the breakdown rate to just one-tenth its former level.

MINIMUM IC/RC DESIGN (ICR-RCR DESIGN)

This section focuses on the LCC design approach, which seeks a balanced, simultaneous reduction of initial costs (IC) and running costs (RC).

Approach and General Procedure for ICR-RCR Design

This approach is most effective in large-scale projects involving considerable investment that aim to use LCC design for entire lines of high-running cost equipment and or multi-unit equipment.

The design procedure is shown in Figure 4-17. The center of the figure shows the LCC design procedure, the left side shows the standards and other data required for the design processes in this procedure, and the right side shows the evaluation methods and the issues under evaluation for each step. The LCC design procedure can be described briefly as follows.

Establish the design mission. The design mission described earlier includes the established LCC target values that define DTC (design-to-cost) as well as other factors listed in the right-hand column of Figure 4-17 (process quality, production capacity or volume, years of equipment use, and operations staff).

Generate the preliminary design (baseline alternative). Once you have established the design mission, study fabrication methods and apply accumulated technology and other technical data to form a preliminary design centered on basic functions (such as quality and mass-production capacity). Various mission requirements must still be met. At this point, the design is simply the base from which to work out more specific design proposals that meet the various needs of the mission. This basic design plan is the baseline alternative. Baseline alternatives help locate the gaps between design plans and mission requirements.

In LCC design, designers evaluate and clarify the needs for both initial and running cost reduction, for example, by identifying the initial cost and running cost items that have the highest contribution ratios. The process is broken down into the following six steps.

Step 1: ICR design. Here you pinpoint and analyze the equipment functions while referring to the baseline alternative. Estimating the initial costs needed for those functions and using the checklist designed to root out improvement needs, you begin planning improvements for reducing these initial costs (minimum IC design).

This method is called function needs (FN) analysis (see pp. 166-177). Use the improvement ideas generated here in drafting several specific alternate plans. At this point also draft an ICR plan and in so doing, list any issues related to running cost reduction that might arise.

Step 2: Estimate LCC. Next, estimate the initial cost and running cost of various cost items related to the design mission. Reductions in initial cost tend to create corresponding increases in running cost.

Step 3: Clarify RCR design needs. With running cost estimates for each function, look for running cost reduction needs again, using the function-needs analysis method.

Step 4: RCR design. Addressing the comments and considerations raised in step 3, work out various alternate plans.

Step 5: Estimate additional initial costs for RCR alternative plans and effects. Compare the alternate plans for running cost reduction developed in step 4 with those for initial cost reduction to estimate additional initial costs and also the effects of running cost reduction.

Step 6: Tradeoffs. At this point, evaluate the various alternate plans from step 5 in terms of engineering economics and select the optimal plan. Base your selection on each plan's economy and on intangible factors such as safety, future use, risks, and so on.

When dealing with particularly large and complex equipment or devices, it may be impossible to come up with a plan that reduces both IC and RC in the six steps above. In such cases, follow the same procedure by first minimizing IC. After selecting an alternate plan that best minimizes initial cost, however, you should adopt that plan and use it as a basis for reducing running cost. This two-step process may still be difficult, but it should help create an innovative ICR-RCR design plan.

Case Study 4-3: ICR-RCR Design

This section illustrates the main points of ICR-RCR design with a simple example.

Background

A company planned to build a new manufacturing line as part of its switch to a new product. Since the new product's specifications were different from those of the current product, process planning called for a new manufacturing line with nine new processes. Since the current profit had already reached its ceiling, the product cost planning needed to be as strict as possible. Therefore, they established a tight budget framework for estimated equipment investment costs (IC) and set detailed target values (as part of the RC mission) for operating costs (RC) based on the proposed process plans. This was the situation the company faced when it launched its LCC design efforts.

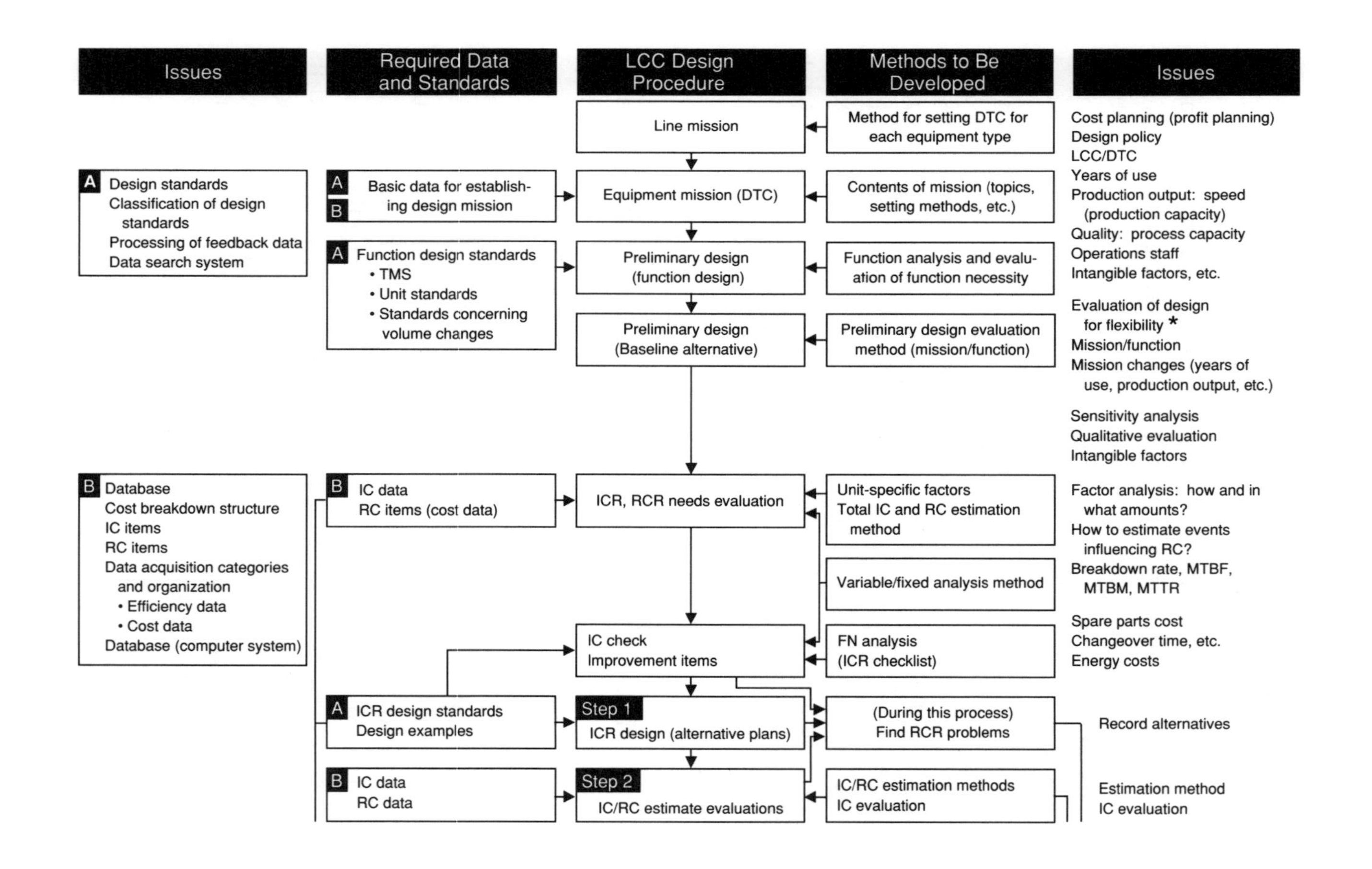
Issues
Required Data and Standards
LCC Design Procedure
Methods to Be Developed
Issues
Line mission
Method for setting DTC for each equipment type
Cost planning (profit planning)
Design policy
LCC/DTC
Years of use
Production output: speed (production capacity)
Quality: process capacity
Operations staff
Intangible factors, etc.
A Design standards
Classification of design standards
Processing of feedback data
Data search system
A B Basic data for establishing design mission
Equipment mission (DTC)
Contents of mission (topics, setting methods, etc.)
A Function design standards
• TMS
• Unit standards
• Standards concerning volume changes
Preliminary design (function design)
Function analysis and evaluation of function necessity
Preliminary design (Baseline alternative)
Preliminary design evaluation method (mission/function)
Evaluation of design for flexibility *
Mission/function
Mission changes (years of use, production output, etc.)
Sensitivity analysis
Qualitative evaluation
Intangible factors
B Database
Cost breakdown structure
IC items
RC items
Data acquisition categories and organization
• Efficiency data
• Cost data
Database (computer system)
B IC data
RC items (cost data)
ICR, RCR needs evaluation
Unit-specific factors
Total IC and RC estimation method
Factor analysis: how and in what amounts?
How to estimate events influencing RC?
Breakdown rate, MTBF, MTBM, MTTR
Variable/fixed analysis method
IC check
Improvement items
FN analysis
(ICR checklist)
Spare parts cost
Changeover time, etc.
Energy costs
A ICR design standards
Design examples
Step 1
ICR design (alternative plans)
(During this process)
Find RCR problems
Record alternatives
B IC data
RC data
Step 2
IC/RC estimate evaluations
IC/RC estimation methods
IC evaluation
Estimation method
IC evaluation

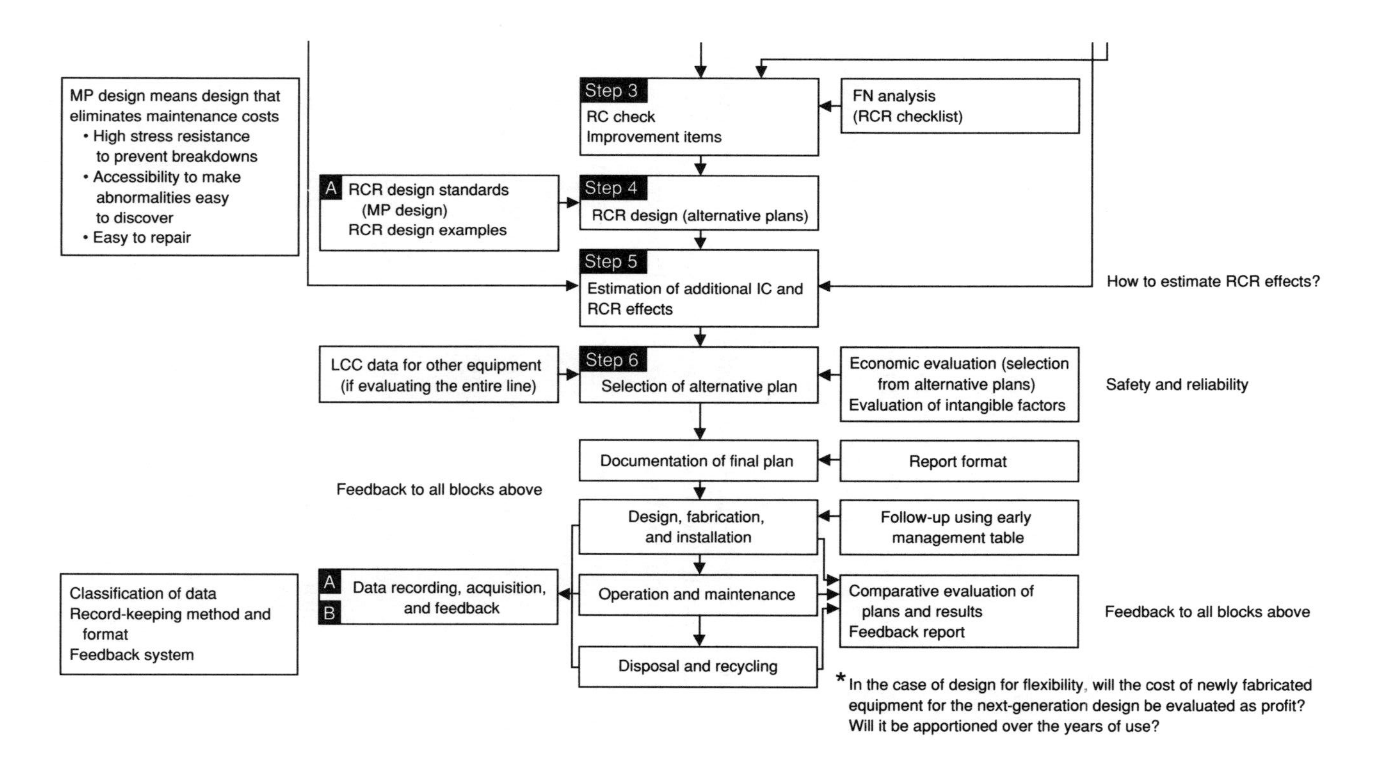

Figure 4-17. Detailed Procedure for ICR-RCR Design

Outline and Structure of LCC Design

Target cost items and target values. Since the equipment investment budget included many new processes, the team established a $1.1 million budget for the entire nine-process line. Based on the current product's running and planning costs, the planners established target costs for labor hours directly related to operations and energy, maintenance, and materials costs. Their total costs came to about $100,000 less than the total estimated in the baseline alternative.

LCC estimation method. The initial cost estimate was based on estimates from the company commissioned to manufacture the equipment. The running cost estimate was based on in-house standard costs (per-unit labor, energy, and materials costs).

Evaluation of LCC suitability. The minimum requirement is to clear the design-to-life-cycle cost (DTLCC). An engineering economics approach was taken to select a plan that exceeds the DTLCC, with a calculated profit margin of 15 percent.

The LCC design procedure described in the previous section was adopted as most appropriate.

Evaluation items for intangible factors. Qualitative judgments were made on the following items:

- Safety
- Reliability (resistance to breakdowns and defects)
- Feasibility of predicted effects
- Accumulation of technology (adoption of new fabrication methods, and so on)

Implementation of LCC design

This example of LCC design implementation is based on the data shown in Table 4-3 and Figure 4-18.

Table 4-3. LCC Implementation Example

Process \ Plan		Plan 0		Plan 1			Plan 2			Plan 3		
		Individual	Total	Indiv'l	Add'l	ROR	Ind'l	Add'l	ROR	Ind'l	Add'l	ROR
Process 1	I	10,500	10,500	35,000	24,500	100% or more	42,000	7,000	100% or more			
	R				67,100			9,200				
	Record	Manual feed operations		Automatic			Additional substation					
Process 2	I	83,900	94,400	99,300	15,700	100% or more	134,600	35,000	12%	174,800	40,200	10% or less
	R				27,000			7,500			2,800	
	Record	Remodeling of existing equip.		Rotor device installed to prevent entanglement			Shape correction device installed			Scale conveyor installed		
Process 3	I	189,500	283,900	244,800	55,200	26%						
	R				14,500							
	Record	New		Automated changeover								
Process 4	I	104,900	388,800	174,800	69,900	35%	265,700	90,900	14%			
	R				28,400			20,600				
	Record											
Process 5	I	279,700	668,500	345,500	65,700	11%						
	R				14,000							
	Record											
	I	90,900	794,400									
Process 8	I	35,000	829,400									
Process 9	I	42,000	871,300									

* Figures shown are in dollars

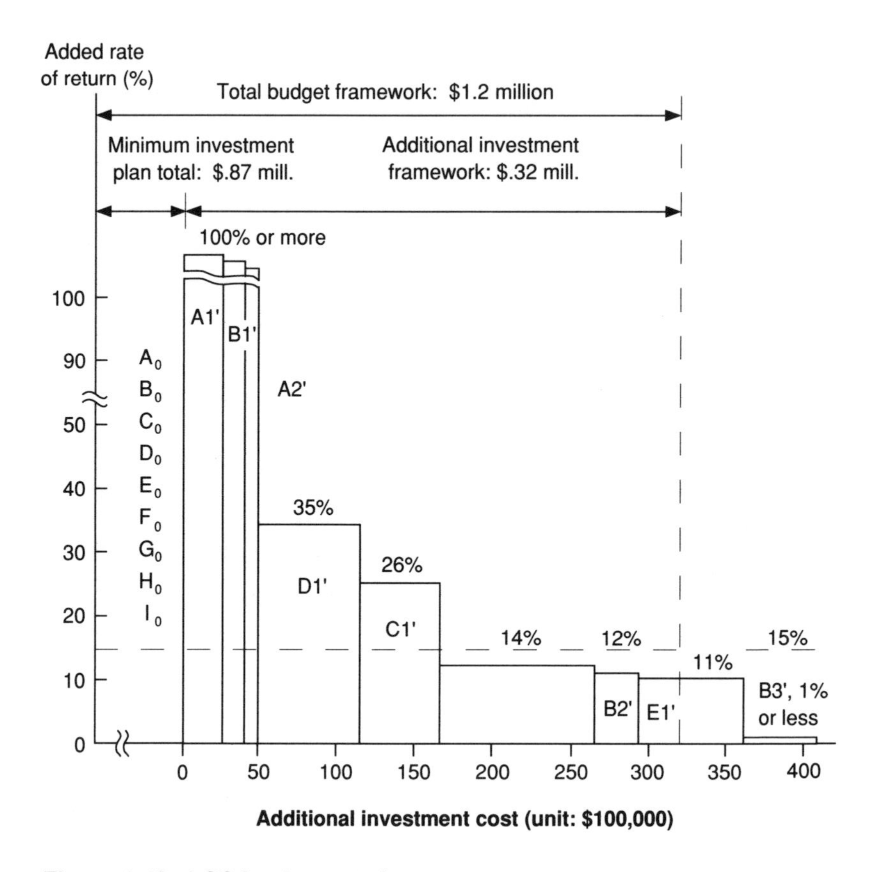

Figure 4-18. LCC Implementation

Step 1: ICR design. The plan 0 column at the left of Table 4-3 shows the ICR design results. This is a realistic plan that establishes a low initial cost for each of the nine processes in the line. If the team were to select plan 0 for all those processes, however, they would end up with high running costs. (In this example, plan 0 is the baseline alternative.)

Step 2: Estimate LCC. At this stage, the team estimated the cost items within the mission (including the running cost targets described above) set for the equipment. They also confirmed that other cost items did not increase much. (Precise calculations for all items are not necessary.)

Step 3: Clarify RCR design needs. Looking at the gaps between the estimates and the target values, the team tried to find as many RC improvement points as possible.

Step 4: RCR design. Using the improvement points from step 3, the team drafted a more detailed specific design plan showing how the design in plan 0 could be improved. For example, under plan 0 for process 1 in Table 4-3, the material feed operation is to be done manually. Since this incurs a large number of labor hours in direct operations, the team should consider automating the material feed operation.

Step 5: Estimate additional IC and RCR. If the running cost improvement plan from step 4 were implemented, what additional initial costs relative to plan 0 would be incurred? How much of a decrease in running cost would it bring?

Refer again to Table 4-3. If it costs about $35,000 in equipment investment to automate process 1, an additional $24,500 will be required, as shown in the "additional" column. This is more than the $10,500 budgeted under plan 0. Plan 1 also provides a reduction in labor hours because of automation, however, the effect of which is to save about $67,100, also shown under the "additional" column. The idea here is to draft as many RCR improvement plants as possible.

Step 6: Trade-offs: Use engineering economics to select a combined plan from various alternate plans. Table 4-3 shows data from various alternate plans (plan 0 to plan 3), listed in ascending order of IC estimates. At the trade-off step, the team examined the alternate plans to find the optimal combination of plans and processes. (Before this could be done, however, they had to eliminate the plans that were not economically feasible; see the discussion of engineering economics on pages 33-38.)

Suppose that plan 1 were selected for process no. 1, plan 2 for process no. 2, and plan 0 for process no. 3. This leaves the arduous task of calculating the optimum overall LCC design plan from this combination of plans. One good way to accomplish this is by applying the engineering economics principle of additional investment profitability to find which combination of plans is best. The percentages shown in the "rate of return" column of Table 4-3 are the profits earned on the additional investment.

The additional investment value =
annual effect value r where $[M \rightarrow P]_{n}^{r}$

Figure 4-18 shows these profit rates arranged in descending order.

The first process/plan combination to go above the investment limit of about $1.13 million or below the minimum calculated profit rate of 15 percent is D_2. At this point, a purely mathematical calculation of the best combination of processes and plans would result in the following selection: A_2, B_1, C_1, D_1, E_0, F_0, G_0, H_0, I_0 To confirm whether this combination fulfills the design mission targets, the team considered the following.

- Initial cost target (maximum) = about $1.13 million. Combination plan total initial cost estimate = about $826,000.
- After calculating the plan's running cost reduction effects relative to the baseline alternative (plan 0), the team obtained an annual RC figure of about $140,000, which is well within limits.

Evaluation of Intangible Factors and Final Effects

Even though the D_2 (process 4, plan 2) combination is one percentage point less than the minimum profit rate, its low running cost makes it hard to dismiss. Therefore, the engineers reworked their RC reduction study based on a minimum profit rate of 14 percent instead of 15 percent, and they took intangible factors (such as equipment reliability, accumulation of technology, and so on) into consideration. The results convinced them that D_2 was worth adopting.

The team also reconsidered the B_2 combination, which includes new equipment that could help reduce the defect rate. Even when the lower defect rate is figured in, however, the low profit rate of 12 percent and the intangible factor evaluation (which highlighted the equipment's uncertain reliability) made this combination unacceptable.

In the E_1 combination, the team not only encountered a low profit rate (11 percent), they were also doubtful about the combination's improvement effect on operating costs. They did not adopt this combination either.

Case Study 4-4: ICR-RCR Design for Painting Unit At Toyota Auto Body

This rather detailed example of ICR-RCR design involves painting units in automobile plants, which are especially appropriate examples of equipment that involves many kinds of investments and costs under both initial and running cost categories. This is precisely why the ICR-RCR design approach is the best for this situation. Thorough implementation of the systematic approach detailed here can be extremely beneficial.

Outline of Issues

Toyota Auto Body had long been working to boost automation in its factory and needed to upgrade its automated equipment in the context of various restrictive conditions. In particular, its upper-coat painting unit needed a wide range of assured characteristics, which in turn required a number of supporting functions that could easily diminish the overall rationalization effect of automation. To maximize rationalization, they had to lower the equipment's total cost (IC plus RC). After evaluating all the cost factors contributing to the upper-coat painting machine's total cost, they carried out an ICR-RCR design approach that emphasized the most expensive cost factor — the paint supply system.

Analysis of ICR

Analyzing initial cost reduction involves the following three steps.

1. ***Review existing equipment.*** ICR points are easier to find if you begin by diagramming the existing equipment (or systems), as shown in Figure 4-19. You should also list the equipment components according to their functions (see the FN analysis table shown in Table 4-4).

2. ***Generate ideas and organize proposals.*** Generate ideas by comparing items shown on the checklist of machine parts (see the *N* side of Table 4-4). Then list the selected ideas in a form like that in Table 4-5. Note also the initial cost and whatever problems have been predicted (under the "abnormalities" column). Also study whatever countermeasures have been suggested and calculate the difference in the running cost.

If it is assumed these proposals will be adopted, they can be organized into independent proposals and mutually exclusive sets of proposals, as shown in Figure 4-20.

3. ***Calculate improvement plan costs.*** First, calculate running cost and initial cost totals for each of the improvement plan combinations mentioned above (see Figure 4-21). Use the following formula:

$$\text{Total cost} = \text{IC} + \text{annual RC} \times [\text{M} \rightarrow \text{P}]_{\text{n}}^{\text{i}}$$

where $n = 9$ years and $i = 10$ percent

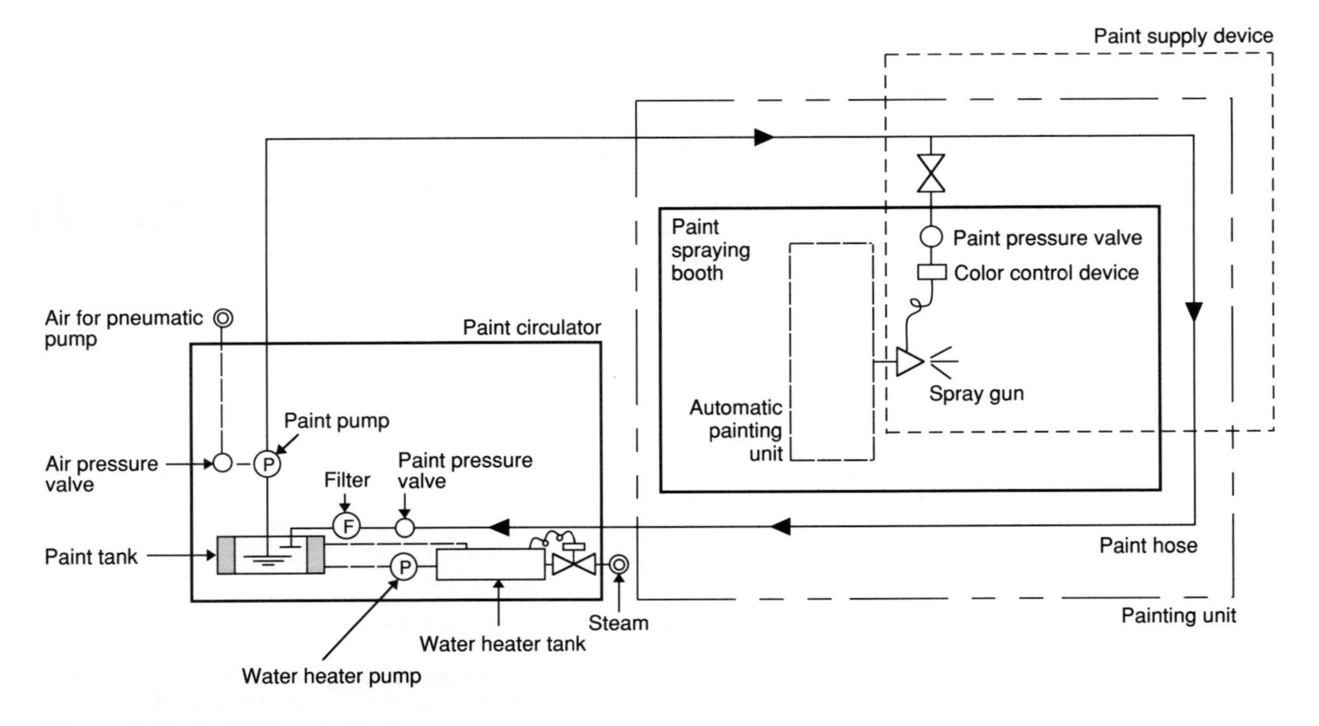

Figure 4-19. Paint Supply System

Table 4-4. FN Analysis Table (for ICR)

Title: Paint Supply System for Automatic Painting Unit

FN Analysis

Category no. ______
Tracking no. ______
Page ______

F: Function/device/part	Target function:	Stable paint supply									Paint supply varies to match body position							Changing paint colors								
	Required functions:	Stable paint pressure							Cont. paint flow		Variable spray pressure				Detect body position			Select valve opened/closed			Open/close specified valve					
N: IC reduction measures		Pressure control valve	Gear pump	Motor	Drive shaft	Transmission gears	Power switch	Power cord/protecting tube	Tubes and joints	Nozzles	Controller	DA converter	Inverter	Power cord/protecting tube	Rotating transducer	Limit switches	Power cord/protecting tube	Operation switches/displays	Controller	Power cord/protecting tube	Paint supply valve	Paint thinner valve	Manifold	Air tubes/joints	Solenoid valve	Power cord/protecting tube
	Step:	1	2	3	4	5	6	7	8	9	10	11	12	13	14	15	16	17	18	19	20	21	22	23	24	25
Common issues	1 Functions vaguely defined?	①			①				Confine pressure adjustment function to pressure valve only																	
Eliminate this part?	2 Any duplicate functions?				①						①															
	3 Replace manual operations?				Visual inspection meter, manually remote controlled																					
	4																									

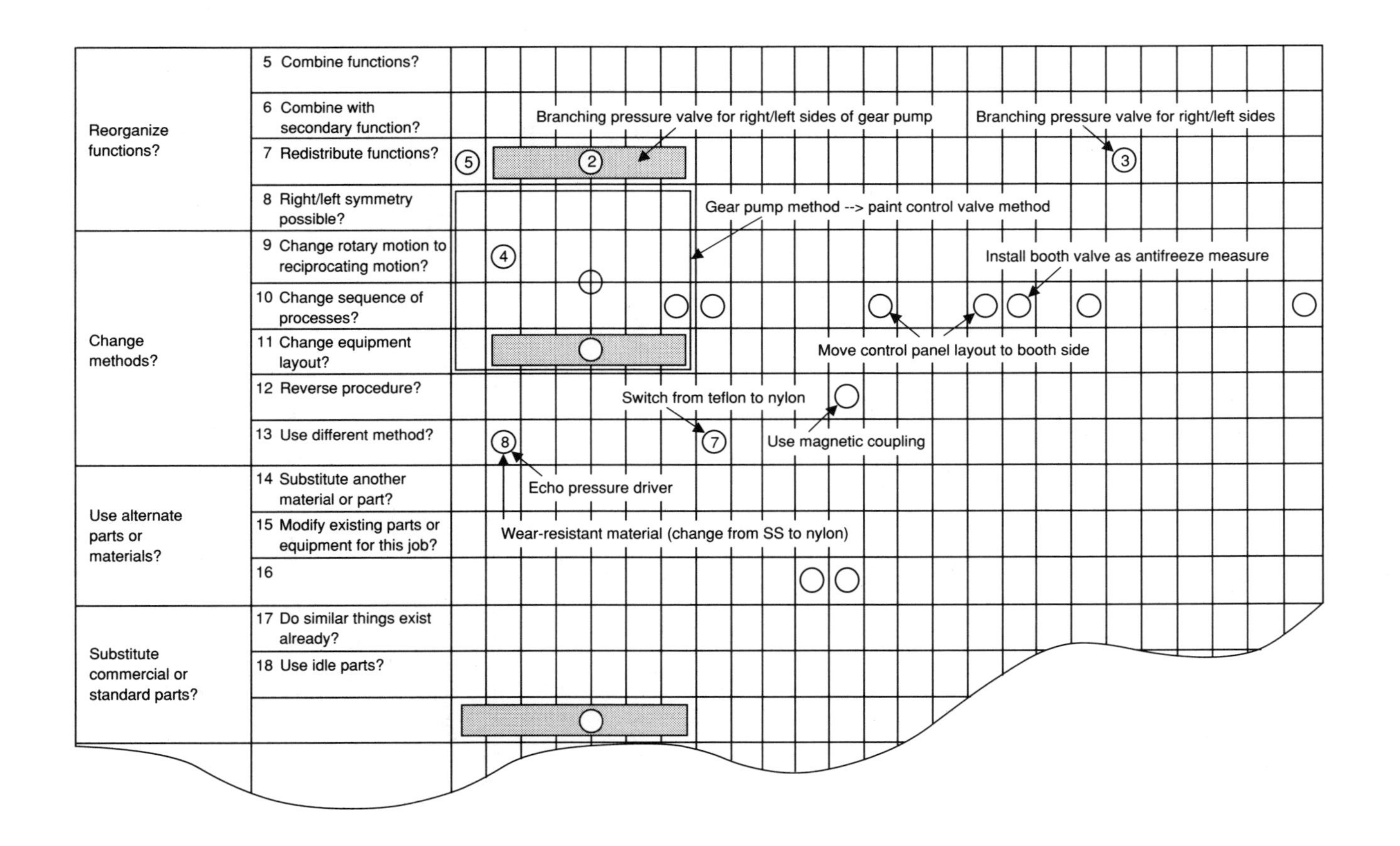

Reorganize functions?
5 Combine functions?
6 Combine with secondary function?
7 Redistribute functions?
8 Right/left symmetry possible?
Change methods?
9 Change rotary motion to reciprocating motion?
10 Change sequence of processes?
11 Change equipment layout?
12 Reverse procedure?
13 Use different method?
Use alternate parts or materials?
14 Substitute another material or part?
15 Modify existing parts or equipment for this job?
16
Substitute commercial or standard parts?
17 Do similar things exist already?
18 Use idle parts?
Branching pressure valve for right/left sides of gear pump
Branching pressure valve for right/left sides
Gear pump method --> paint control valve method
Install booth valve as antifreeze measure
Move control panel layout to booth side
Switch from teflon to nylon
Use magnetic coupling
Echo pressure driver
Wear-resistant material (change from SS to nylon)

Table 4-5. IC Reduction Implementation Items

IC Reduction Item	ICR Effect	Abnormalities	Counter-measures	RC Difference	LCC
1. Gear pump system ↓ Paint regulator system	Current: 53.2 ▲18.6 34.6	• Occur in steady stream • Wide variation	• Manual adjustments • Manual adjustment using regulator	87.8 ▲44.9 +120.4 163.3	198.0
2. R/L unification for gear pump unit	Current: 53.2 ▲9.6 43.6	• R/L cannot be controlled independently	• Insufficiently covered areas can be hand-sprayed	87.8 ▲22.4 +153.6 219.0	262.6

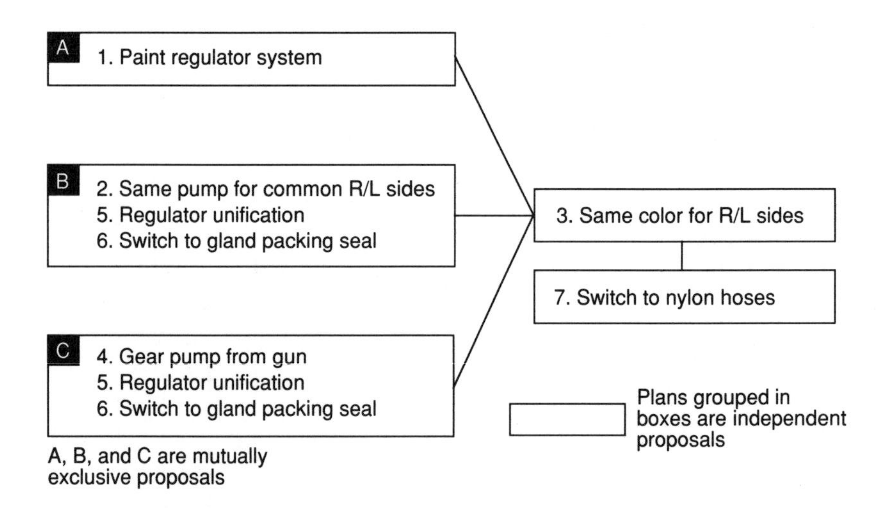

Figure 4-20. Plan Combination Chart

These calculations show that plan A is the most economical. Therefore, plan A is the basis for the next step, running cost reduction (RCR).

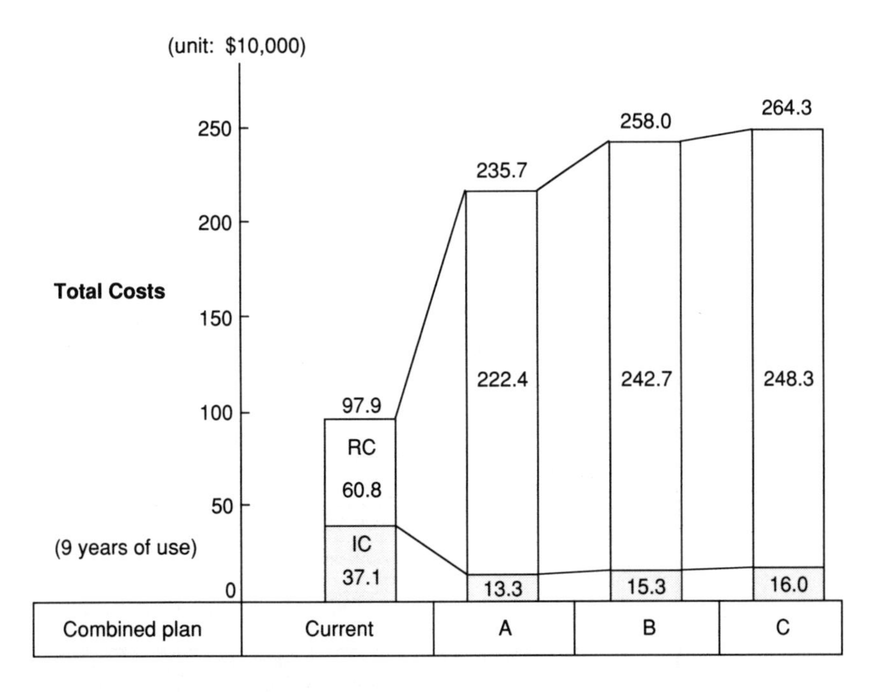

Figure 4-21. Calculating Total Costs for IC and RC

Running Cost Reduction Study

Organize data to facilitate study. Lay out the improvement plans in a table to make the ICR issues easier to see. Then arrange the components into function categories, as shown in the *F* side of the FN analysis table in Table 4-6. On the same table, list the RCR improvement needs that correspond to the ICR improvement plans shown on the *N* side of the FN analysis table.

Select ideas for evaluation and further study. Next, select ideas from the ICR improvement plans and RCR improvement needs listed in Table 4-6. Then estimate the additional ICR and

Table 4-6. FN Analysis Table (for RCR)

Title: Paint Supply System for Automatic Painting Unit

Item FN analysis

Category no. ____
Tracking no. ____
Page ____

F Function device part	Target function:	Stable paint supply									Paint supply varies to match body position							Changing paint colors								
	Required functions:	Spray pressure that varies with paint temperature							Cont. paint flow		Press gauge	Variable spray pressure			Detect body position			Select valve opened/ closed			Open/close specified valve					
N RC reduction measures		Paint regulator	Paint pressure gauge	Remote adjust valve	Thermometers				Tubes and joints	Nozzles	Controller	DA converter	Inverter	Power cord/protecting tube	Rotating transducer	Limit switches	Power cord/protecting tube	Operation switches/displays	Controller	Power cord/protecting tube	Paint supply valve	Paint thinner valve	Manifold	Air tubes/joints	Solenoid valve	Power cord/protecting tube
	Step:	1	2	3	4	5	6	7	8	9	10	11	12	13	14	15	16	17	18	19	20	21	22	23	24	25
Reduce operation time	1 Improve control precision	(2)																								
Reduce repetition in operation procedures	2	Add'l paint temp. control or control in glass																								
	3																									
	4																									

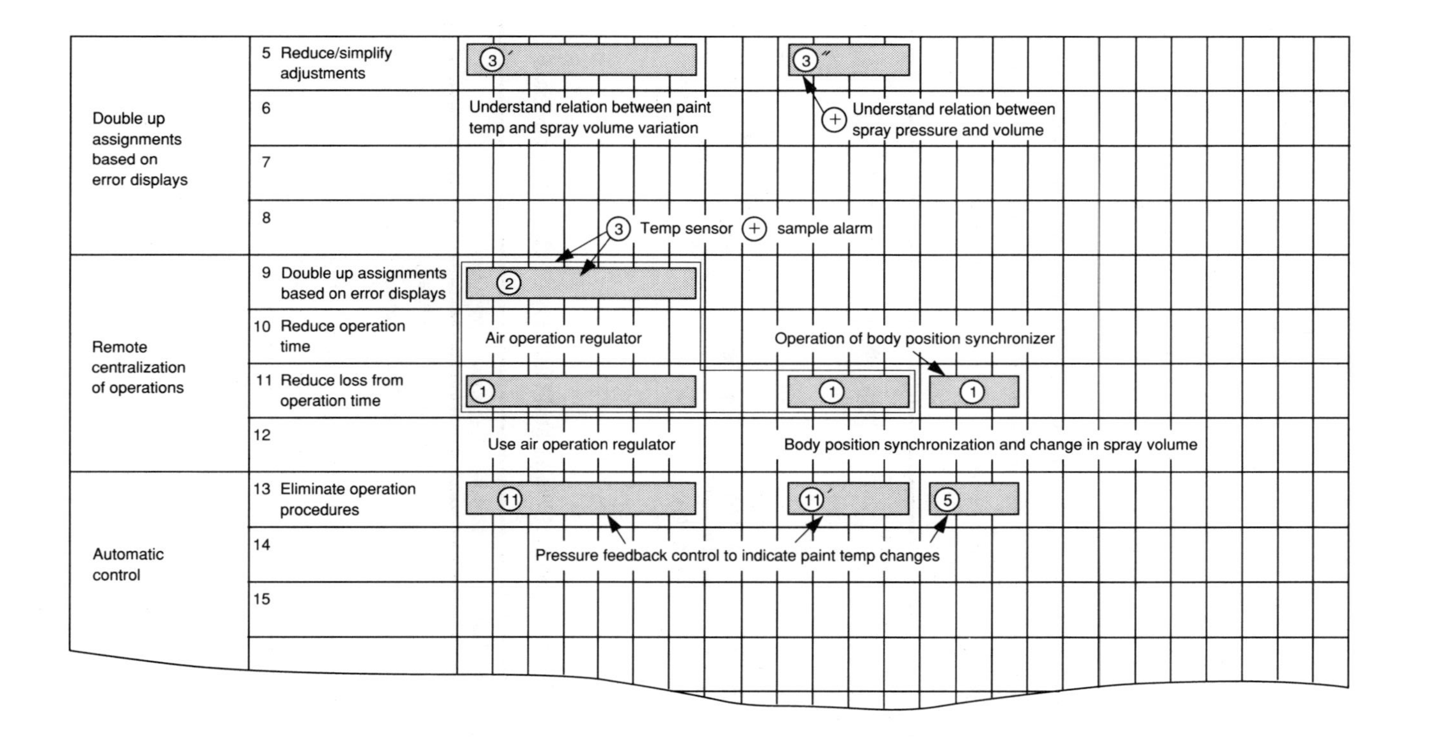
Double up assignments based on error displays
Remote centralization of operations
Automatic control
5 Reduce/simplify adjustments
6
7
8
9 Double up assignments based on error displays
10 Reduce operation time
11 Reduce loss from operation time
12
13 Eliminate operation procedures
14
15
3′
3″
Understand relation between paint temp and spray volume variation
Understand relation between spray pressure and volume
3 Temp sensor + sample alarm
2
Air operation regulator
Operation of body position synchronizer
1
1
1
Use air operation regulator
Body position synchronization and change in spray volume
11
11′
5
Pressure feedback control to indicate paint temp changes

RCR effects these ideas will have if they are used. Organize these data (see Table 4-7) and list any ideas that cannot be implemented using existing technology as R&D issues.

In Table 4-8 independent and mutually exclusive sets of proposals are organized based on the earlier organization of the same proposals for initial cost reduction (see Figure 4-20).

Table 4-7. ICR and RCR Effects of Selected Improvements

Abnormalities (RCR improvement points)	Can be implemented using existing technology	IC	RC	Development issues	Predicted Values	
					IC	RC
1. Cannot change spray volume. Occurs steadily	Automatic adjustment of pressure at nozzle	14.5	▲51.0	———		
3. Significant deviations. Occurs steadily	Visual inspection and adjustment of flow rate gauge	1.5	▲36.5	• Paint temperature and pressure, feedback control	8.0	▲42.0

Table 4-8. Independent Plans and Mutually Exclusive Proposals

Independent Plan	Mutually Exclusive Proposals
Plan 1	1-1 Install manual remote adjustment device 1-2 Install paint temperature visual inspection gauge and alarm device 1-3 Combine Proposals 1-1 and 1-2 1-4 Install paint temperature control
Plan 3	3-1 Install one-way valve 3-2 Use one-way valves as paint valves
Plan 7	7-1 Use teflon for paint hoses (hoses after paint valves)

Selection of Improvement Plan

At this point, you must select from among the combined proposals listed as mutually exclusive under each independent plan (Table 4-8).

Figure 4-22 shows a graph with the RCR effects of the selected ICR improvement plan (plan 1) as the *Y* axis and the same plan's additional IC as the *X* axis. The dotted line in the figure shows the reduction value *r* (10 percent) of the current plan. The solid lines show the reduction values of various mutually exclusive sets of proposals. The plan that most exceeds the value of the dotted line at any parallel point along the *Y* axis is the optimum plan (see also Table 4-9). If, however, all the RCR improvement plans fall below the $r = 10$ percent average for the current plan, you should select the current plan.

Review Process

The following figures illustrate the subsequent evaluation of the selected improvement plan and development concepts aimed at further improving the RC.

Evaluation of improvement plan. (See Figure 4-23.) While this plan has significantly lower and initial running costs (including considerably reduced maintenance costs), paint costs would be greatly increased.

Problems and R&D issues in improvement plan. (See Figure 4-24.) This document highlights a technological issue in an improvement plan: paint spray volume fluctuates too much when manual adjustments are made. There is a need to automate temperature detection and paint pressure adjustment.

Development concepts. (See Figure 4-25.) Here the problem highlighted in Figure 4-24 is addressed by a specific proposal for automating temperature detection and paint pressure.

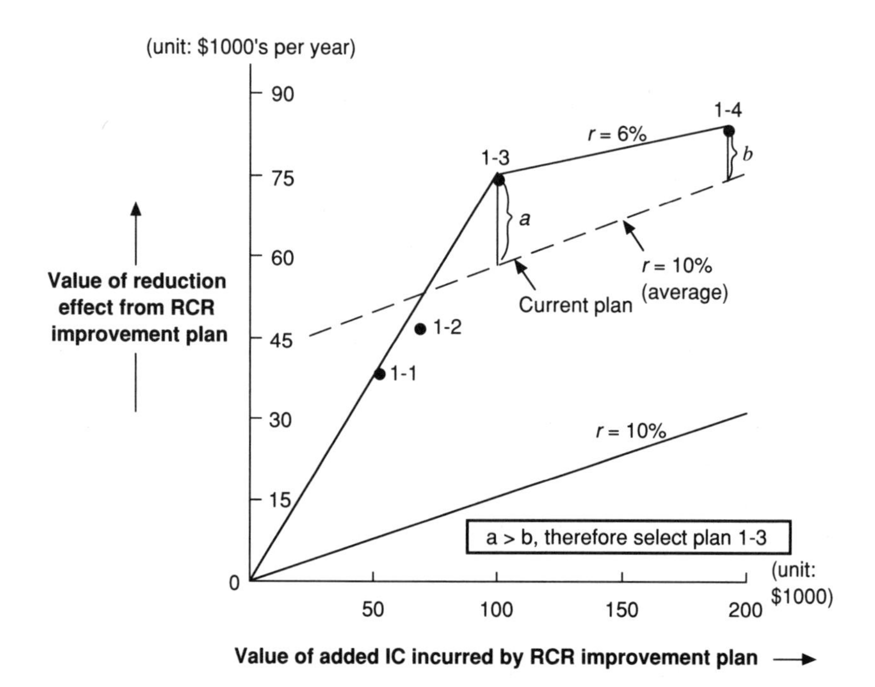

Figure 4-22. RCR Improvements for ICR Improvement Plan (Plan 1)

Table 4-9. Independent Plans and Mutually Exclusive Proposals

Independent Plan	Mutually Exclusive Proposals	Proposals Selected
Plan 1	1-1 Install manual remote adjustment device 1-2 Install paint temperature visual inspection gauge and alarm device 1-3 Combine Proposals 1-1 and 1-2 1-4 Install paint temperature control	Proposal 1-3
Plan 3	3-1 Install one-way valve 3-2 Use one-way valves as paint valves	Current plan
Plan 7	7-1 Use teflon for paint hoses (hoses after paint valves)	Proposal 7-1

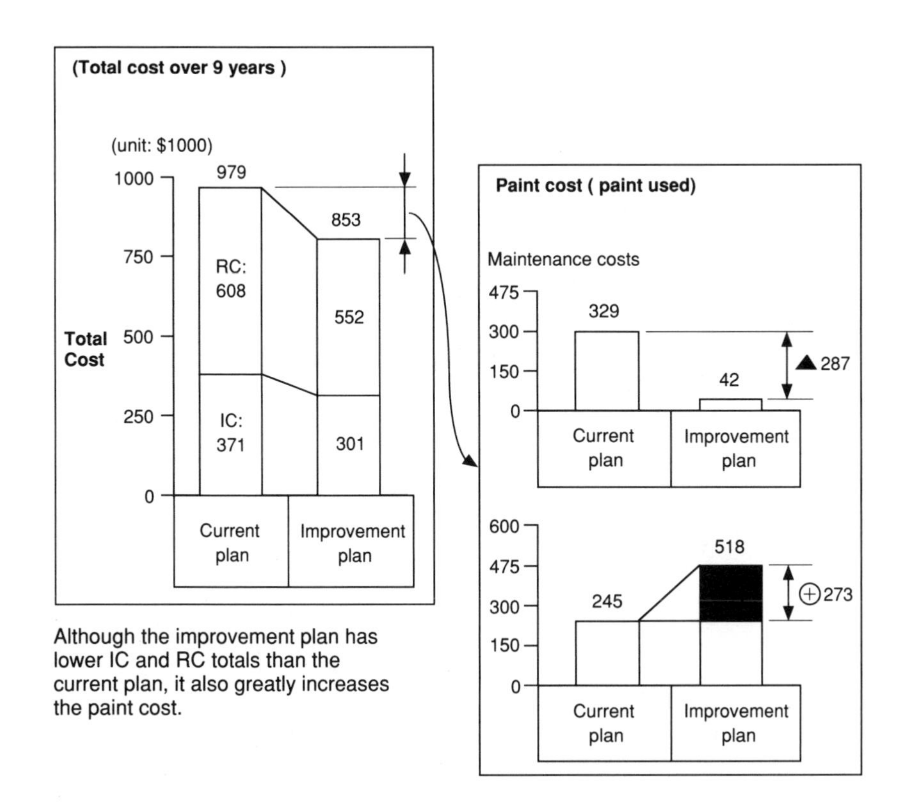

Figure 4-23. Evaluation of Improvement Plans

Estimated effects after development. (See Figure 4- 26.) This graph displays the projected benefits of developing the technological solution proposed in Figure 4-25.

LCC DESIGN UNDER UNCERTAIN CONDITIONS (DESIGN USING SENSITIVITY ANALYSIS)

Design using sensitivity analysis (SA) is another approach to LCC design. SA is used to help select the most economical design plan in terms of LCC. According to the Japan Institute for Plant Maintenance, sensitivity analysis is a method for analyzing

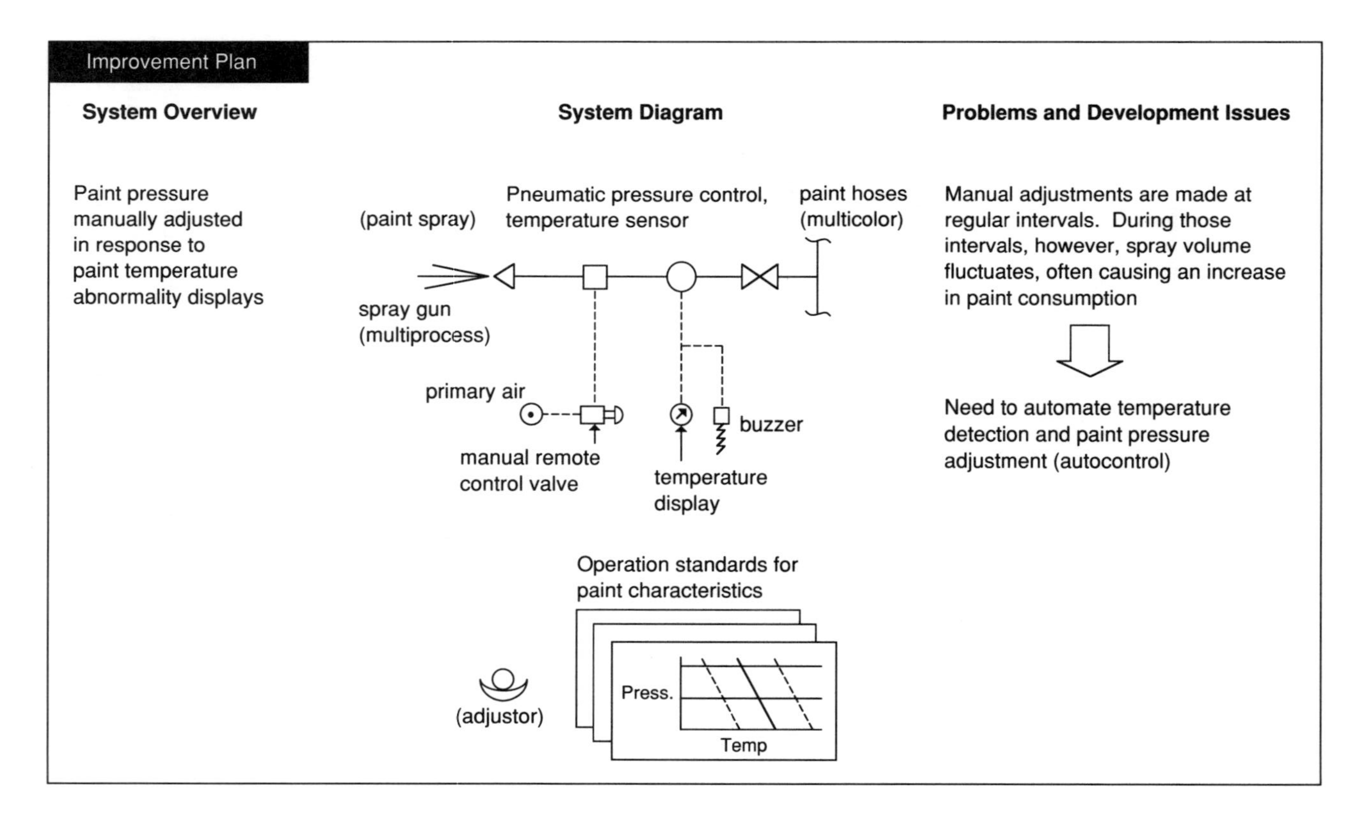

Figure 4-24. Problems and R & D Issues in Improvement Plan

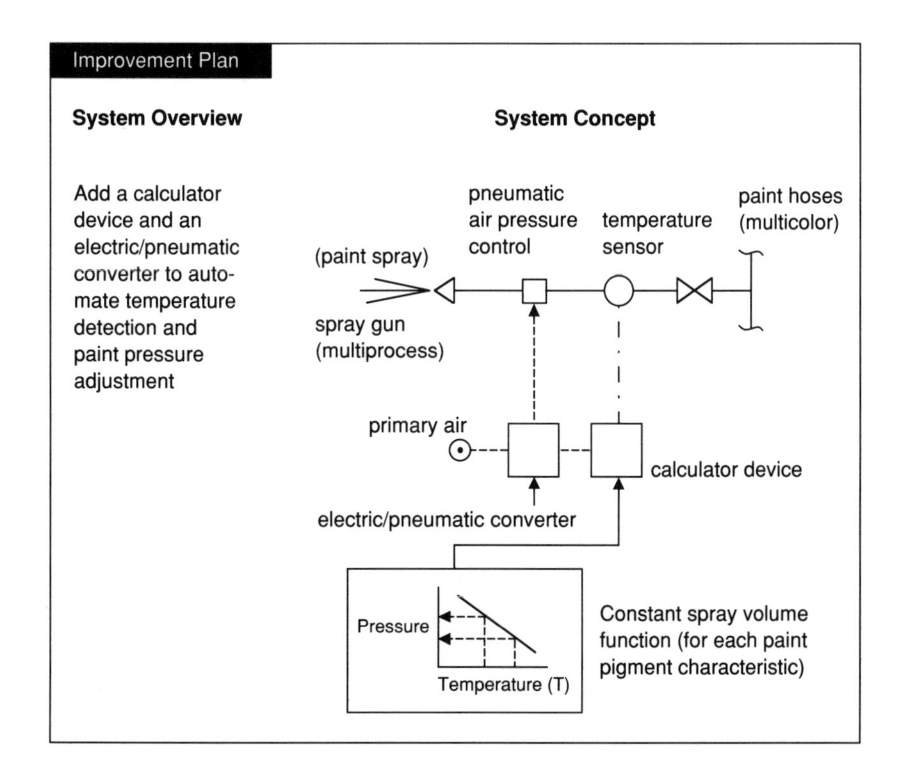

Figure 4-25. Development Concept

a project's safety in cases where the equipment investment plan must be analyzed using uncertain data and where the probability distribution of the project's variables cannot be determined.

Consider using sensitivity analysis as part of your LCC design method when one of the basic design factors must remain unclear. For example, suppose no predictions can be made about how well a new product will sell until it is on the market. This uncertain factor (sales volume) makes it difficult to know how much equipment investment is needed to support the required production volume. Sensitivity analysis establishes a range for uncertain factors and thus helps in selecting the best LCC design plan for that range.

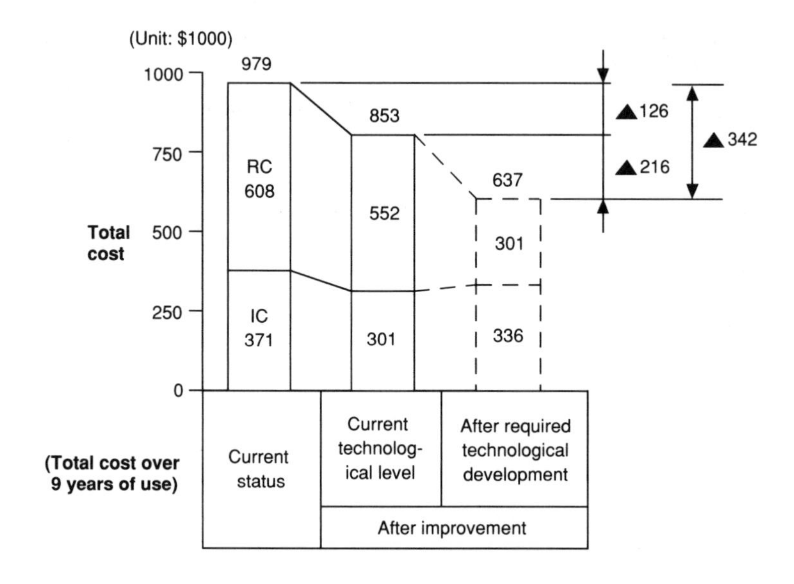

Figure 4-26. Estimated Effects at Completion of Development

All the LCC design approaches described earlier are suitable for cases in which the basic design mission is clear and you are ready to begin designing the equipment in detail. The SA method is well suited for cases in which there are one or more vague formulations of the design mission and in which the LCC values are high. (The latter includes not only high equipment investment costs but also correspondingly high manufacturing costs, where mistakes in planning can lead to large losses.) In such cases, there are two principle issues:

- What fabrication methods should be selected (or developed)?
- Which of the existing equipment plans (or investment plans) should be selected?

The method for answering these questions is generally known as risk analysis. Quantitative estimates must be provided for unknown factors; otherwise, no progress can be made in the logical, measurable world of equipment design.

In design using SA, variance of uncertain factors is handled by asking the following two questions based on preliminary LCC calculations for the design plan under consideration.

- What is the permissible range of variance (in terms of safety, profitability, etc.) for each uncertain factor?
- By how much must the LCC be reduced to produce a design plan that allows for a certain range of variance in its design factors? (In other words, how low must the LCC be to ensure profitability?)

With these questions in mind, consider the following general procedure for design using SA.

1. Create figures and tables to help team members visualize and clarify problems.
2. Investigate the various possibilities and draft plans.
3. Quantify each plan's uncertain factors and their consequences.
4. Think of engineering methods that can reduce the negative impact of major uncertain factors.
5. Select the plan that ensures maximum profitability under worst-case scenarios involving uncertain factors.

Approach and Procedure for Design Using SA

With uncertain factors, there are no logical certainties, and luck plays a role. The following three cases illustrate design using SA, each using a diagrammed model based on engineering economics.

Model 1: Variable Sales and Cost Results

In this method the various elements of the sales and cost results are diagrammed to see how variability in the elements affects profitability as a means of finding the optimal design plan. Figure 4-27 shows one such diagram, which includes a product with unit price a that is to be made using design plan A (at investment cost C and per-product overhead cost v).

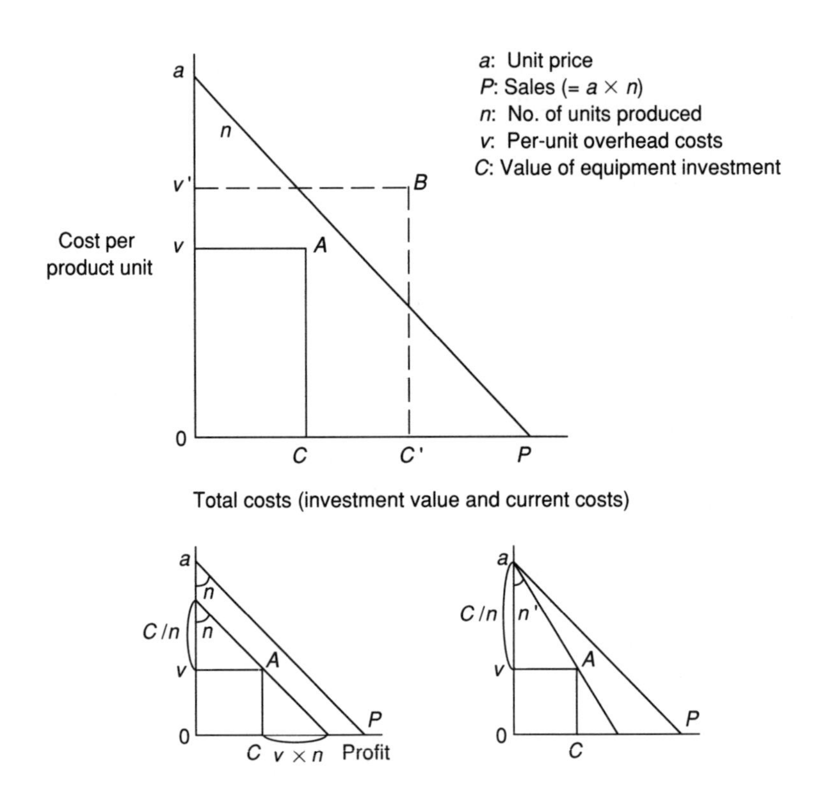

Figure 4-27. Model 1 for Design Using SA: Variable Sales and Cost Results

If a straight line is drawn from a to P via A, with the constant n as the number of products, then

$$\text{Unit cost} = v + \frac{c}{n}$$

$$\text{Per-unit profit} = a - \text{unit cost}$$

$$\text{Total profit} = P - (v \bullet n + C)$$

$$\text{Interest Rate} = \frac{P - (v \bullet n + C)}{v \bullet n + C}$$

You can determine the number of products needed to make the unit price equal to the cost (the angle of n') by using the following formula:

$$n' = \frac{C}{a - v}$$

Plan B in Figure 4-27 (which has investment cost C' and per-product cost v') is a slightly different perspective on the same problem. If a, n, and P are all fixed (the line from a to P does not change), and if the unit cost = the cost, you can calculate the relationship between loss-incurring costs C' and v' as follows:

$$a \geqq v' + \frac{C'}{n}$$

$$v' \leqq -\frac{C'}{n} + a$$

Note that v' and C' meet at a point above the line that connects points a and P. In other words, any plan whose cost elements meet at a point above this line will incur a loss instead of a profit and therefore is not acceptable.

Model 1: Uncertain Production Volume

Another simple example is shown in Figure 4-28, where a = cost target, and C = allowable investment total. Suppose, however, that the production volume is an uncertain factor, with an estimated range between n_1 and n_2.

What kind of problem does this estimated range for n pose?

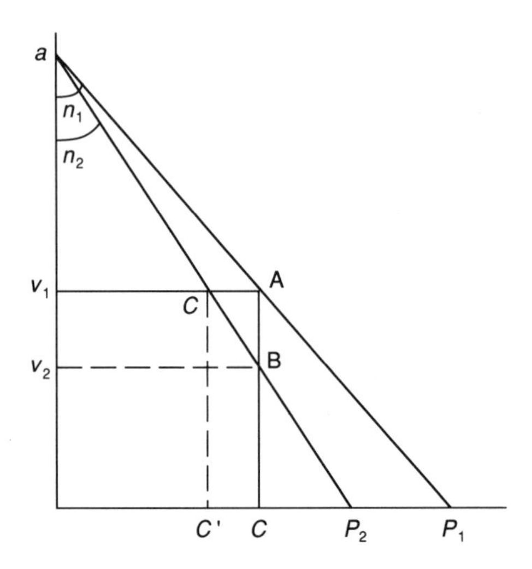

Figure 4-28. Model 1 for Design Using SA: Uncertain Production Volume

In Figure 4-28, investment plan A (C and v_1) will not be profitable even if the production volume reaches the maximum point on the estimated range. Consequently, the following alternate plans might be suggested:

- Plan B: Change from v_1 to v_2.
- Plan C: Change from C to C'.
- Plan D: Start with C' and v_1, and commit additional investment later if warranted by production volume conditions.

Model 2: Uncertain Sales Volume

This method uses a diagram illustrating the profit/loss breakeven point. Figure 4-29 shows two investment plans proposed for a product with unit cost a.

- Plan I: Equipment investment cost = C_I (fixed cost) per-unit cost = v_I (proportional cost)
- Plan II: Equipment investment cost = C_{II}

Per-unit cost = v_{II}

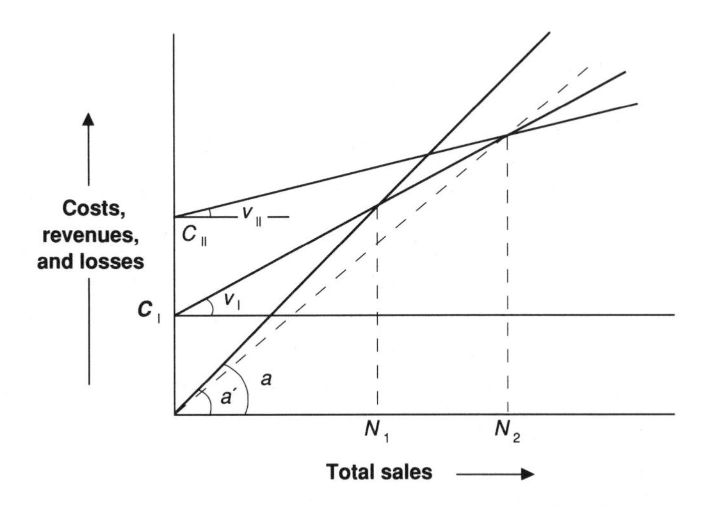

Figure 4-29. Model 2 for Design Using SA: Uncertain Sales Volume

The sales volume is the uncertain factor in this case. Three sales-volume scenarios follow for plan I and plan II, based on the breakeven points shown in Figure 4-29.

- If the sales volume is less than N_1, both plans result in losses.
- If the sales volume is between N_1 and N_2, plan I is more profitable than plan II.
- If the sales volume is at or above N_2, plan II is more profitable than plan I.

You can select the scenario with the best chance of matching the sales estimates. Based on those sales estimates, the options are to

- draft another investment plan
- adopt plan I
- adopt plan II

You might also consider the following questions:

- Can the v_I factor in plan I be reduced slightly?
- Can the v_{II} factor in plan II be reduced slightly?
- What happens if the unit price a is changed?

Suppose, for example, that the company is likely to adopt a policy that emphasizes capturing market share and ensuring a high sales volume by lowering the unit price a. The breakeven point for plan I and plan II will depend on maintaining unit price a' for sales volume N_2. If it is likely that the unit cost can be lowered to a', plan II is the best choice.

Model 3: Uncertain Years in Use and Running Cost

Suppose that three types of machines (shown as A, B, and C in Table 4-10) are technically feasible for processing a certain part. In this case, both the years of use and the annual running costs are uncertain factors. Which of the three types of machines is best suited for this situation? If only the annual costs of the three plans must be compared with investment plans having different years of use, the diagram shown in Figure 4-30 can be used as the statistical tool for comparison purposes.

The horizontal axis shows the amount of initial investment for each plan (shown as A, B, and C). The lines drawn from the origin are plotted according to the capital recovery factor based on interest i and years of use. To find the annual cost of equipment in use for three years with an initial investment cost of $50,000, draw a dotted line vertically from point A until it inter-

Table 4-10. Three Alternate Improvement Plans

Machines under consideration	Estimated years of use	Initial investment	Annual running costs
Machine A (custom)	3	$56,000	$21,000
Machine B (general purpose, compact)	10	$91,000	$38,500
Machine C (general purpose, large)	15	$175,000	$28,000

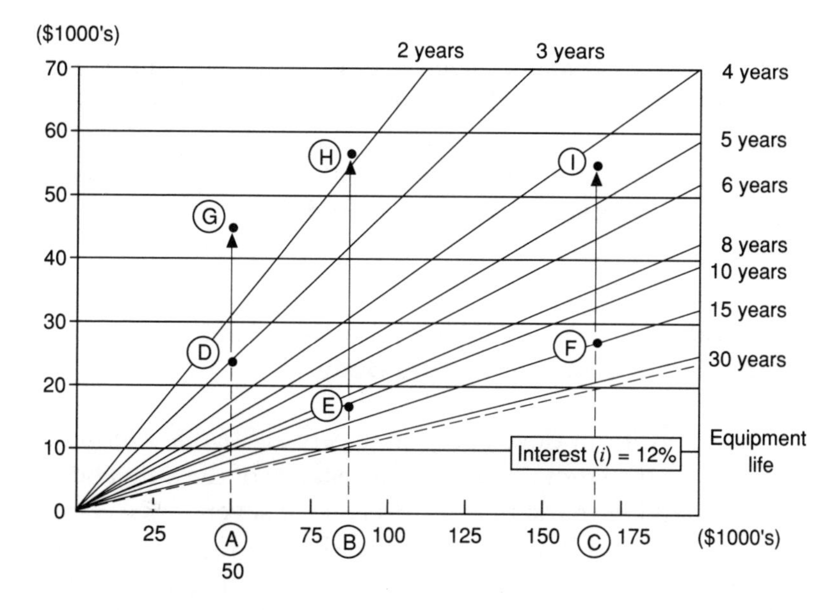

Figure 4-30. Diagram for Comparison of Alternate Investment Plans

sects (at point *D*) the line that connects the origin and the "3 years" point at the top of the figure. Determine the annual cost by finding point *D* on the vertical (annual cost) axis, which in this case corresponds to about $24,000 per year. In the figure, points *D*, *E*, and *F* were all determined this way.

Next, find the total costs by adding the annual running costs to the annual investment cost figures. These total costs are shown as points *F*, *H*, and *I* in Figure 4-30. The lower these points are on the vertical axis, the more economical are the plans they represent. If this analysis is also to include a material cost inflation factor, redo the diagram and replace the nominal interest factor *i* with a real interest factor *k*.

Thus, Figure 4-30 shows a number of things. For example, this diagram for a sensitivity analysis indicates that, even if a model change makes it impossible to use machine A for more than two years, machine A is still preferable to machines B and C. If machine A can be used for at least three years, it is preferable to machines B and C, even if those machines can be used for 30 years. In addition, this figure can be used to examine the effects of alternate measures. Suppose an additional $13-20,000 in initial investment for machine B results in a 40 percent reduction in running costs. Machine B would thus be preferable to machine A.*

Case Study 4-5: QA/SA Design for a New Spot Welding Process at Toyota Auto Body

One priority in developing new cars at Toyota Auto Body is to improve the strength and rust resistance of the auto body to meet diversifying market needs. Because of severe price competition, Toyota Auto Body is also obliged to set tight cost targets. In this instance, to maintain the required welding quality, they needed to automate 13 of the 39 welds in the underbody process, which is the most important process for ensuring auto body strength (see Figure 4-31).

---

* The author thanks Professor Emeritus Shizuo Senju of Keio University for contributing this example.

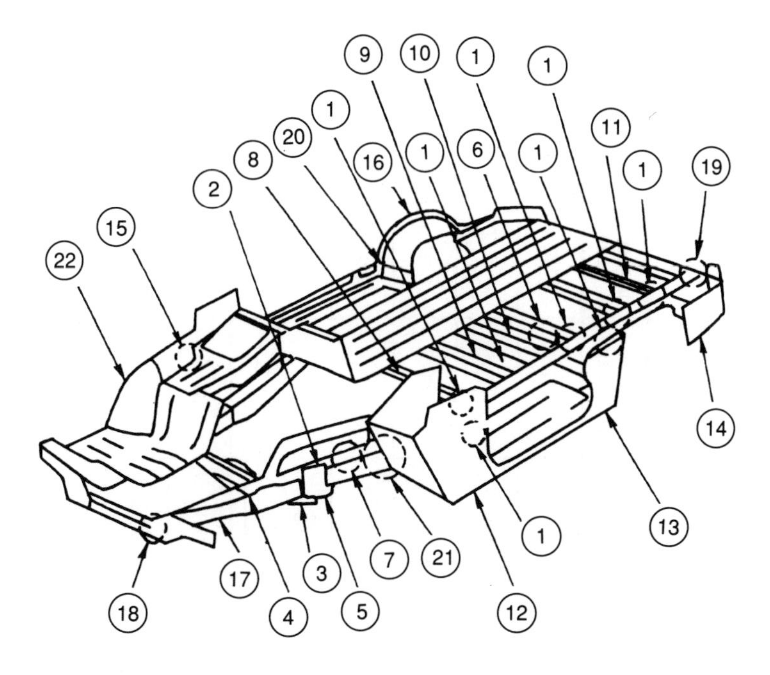

Figure 4-31. Check Points for Automated Spot Welding

Table 4-11 shows the automation method, investment costs, and per-unit value of effects calculated for these automated welds. As shown in the table, this automation plan produces positive per-unit values and meets the quality assurance needs, but it requires a large increase in investment costs for automating the 13 welds. Therefore, the engineering team had to find a cheaper way to provide the required quality assurance without exceeding their strict cost targets.

Economic Analysis

First, the design teams conducted an economic analysis aimed at discovering and improving elements with poor investment efficiency.

Table 4-11. Automation Investment Examination Results

Part	Automation method	Contact points	Labor hours reduced	Per-unit effect value	Investment value	Total cost
1 Cross member ⊕ rear member	Robot X gun		4.33 minutes	$1.31	$196,000	$210,000
2 Front side member ⊗ upper connector	Automated device		0.8	.24	63,000	58,000
3 Suspension member ⊗ upper connector	Automated device		0.8	.23	42,000	26,000
4 Suspension member ⊗ gusset	Biaxial automated device		0.4	.13	24,500	21,000
5 Upper connector AS	Multi		0.3	.09	24,500	24,500
26 Rear floor ⊗ wheel housing	Automated device		0.8	.23	61,500	61,000
27 Apron ⊗ member	Biaxial automated device		0.5	.15	47,000	15,500
28 Front floor AS	Multi		1.3	.40	119,000	147,000

Examination of production volume requirements. To see whether or not production volume requirements could be studied in terms of the planned target volume conditions, the M team checked the production indices from its current mass-production model.

Figure 4-32 shows that model changes in rival automobiles do not have a significant impact on production volume after the third year of production. This led the engineers to propose the following two plans:

- Plan 1: Planned production volume: 6,300 units/month
- Plan 2: Planned production volume: 6,300 units/month for the first two years, 4,400 units/month thereafter

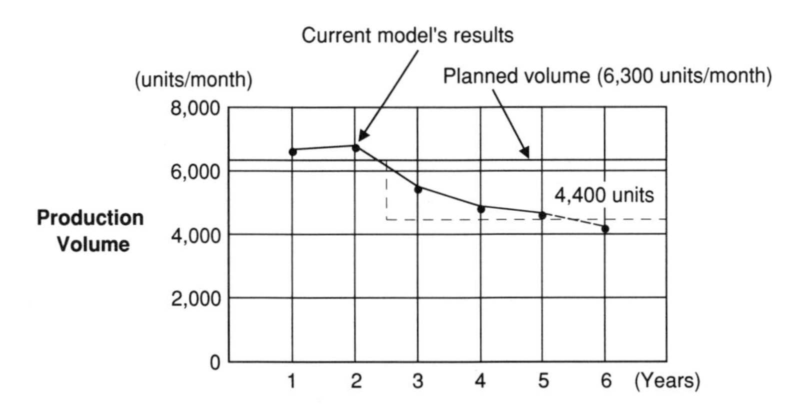

Figure 4-32. Toyota Auto Body's Current Mass Production Model

Selection of improvement targets. The results of an economic analysis are not guaranteed to reveal all of the items needing improvement, so the engineers also drew up a profitability check diagram, shown in Figure 4-33.

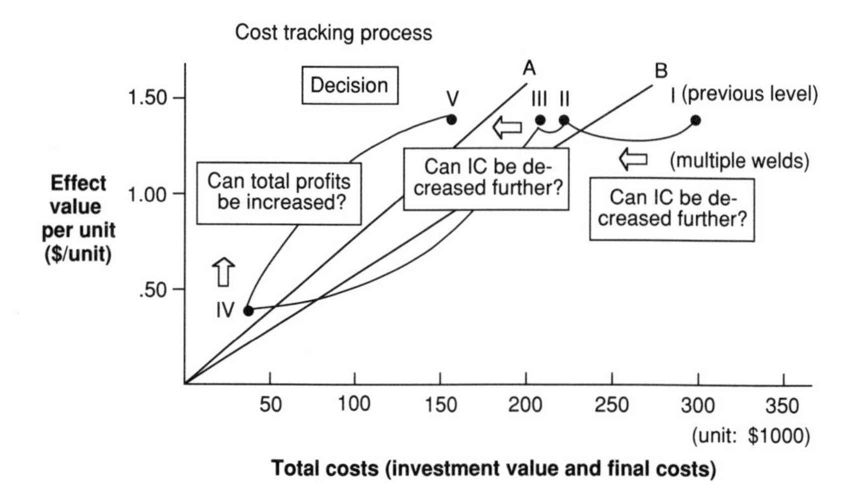

Figure 4-33. Automation Profitability Check Diagram

Lines *A* and *B* are reference lines indicating the allowable range of profitability, based on the following formula:

$$\text{Profit rate} = \frac{\text{Total profit}}{\text{Total cost}} = 100\%$$

This can also be stated as:

$$\text{Efficient amount} = \frac{\text{Invested amount} + \text{total cost}}{\text{Production output}}$$

When an investment plan comes below this line, a 100 percent profit on the targeted rationalization-oriented investment can be realized.

Line *A* also represents production volume N_1 (4,400 units per month) while line *B* represents production volume N_2 (6,300 units per month). The horizontal axis shows the final costs (IC and RC) — the total costs over the life of the model — and the vertical axis shows the per-unit effect value.

The allowable profit reference lines are set at the 100 percent profit level because the company's standard for profit on rationalization-oriented investments is 100 percent or above. Two allowable profit reference lines in the figure provide a check to ensure profits, even when the production volume goes down, and a reference line for planned production volume on the auto body line, where money is being invested for automation.

Either of the following two types of improvements is needed when the estimated profitability drops below the allowable profit reference lines:

- Lowering the total cost while maintaining the per-unit effect value
- Maintaining the total cost while raising the per-unit effect value

Next, the team used a profitability checklist to review the results of the automation-oriented investment examination (Table 4-11). Two types of problems were discovered during this check (see Figure 4-34): problems that hurt profitability (16: Wheel house inner AS, 28: Front floor AS), and problems that do not hurt profitability but increase costs (1: Member AS, 6: Rear floor AS).

Examination of Improvement Plans

At this point, the team used a checklist showing the basic functions of the existing equipment to help them discover how to replace, simplify, or combine functions to reduce auto body processing costs (Figure 4-35). After reviewing this checklist carefully they were able to formulate automation plans that lead to higher profitability. Figure 4-36 is a flowchart of the improvement plan review process. A specific example, the improvements made for the front floor process, will clarify the examination.

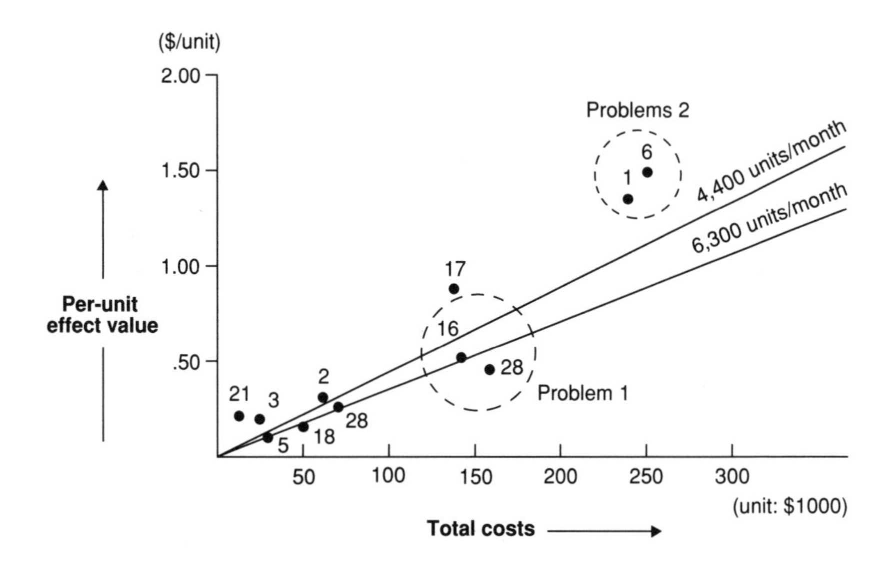

Figure 4-34. Per-unit Effect Value

As shown in Figure 4-37, the front floor process involves the attachment of a small bracket onto a large panel. After cross-checking the functions and review items (using the checklist in Figure 4-35), the engineering team studied parts of the process needing improvement and noted a few points of concern. Next. they drafted alternate proposals focused on these points, reviewed the equipment configuration, and prepared specific improvement plans. Figure 4-38 illustrates this process.

Review Results

Figure 4-39 shows the results of studying the various spot welding sites. As shown in the figure, the team succeeded in improving the per-unit value at each welding site.

SOME FINAL TIPS ABOUT USING LCC DESIGN

The following sections offer practical suggestions for maximizing the potential of LCC design.

Databases

To carry out economic evaluations during the design process, requires some sort of database of running (RC) data. You must establish RC cost items and know what kind of raw data to collect and keep for making cost calculations. The raw data should probably include the following:

- Part-specific or unit-specific MTBF data
- Part-specific or unit-specific breakdown-related downtime, repair labor hours, and other related costs
- Labor hours and other costs for part-specific or unit- specific preventive maintenance
- Defect rate (yield) trends
- Changeover time and labor hours
- Trends for operation-related labor hours, materials, energy per-unit costs, and so on

You will also need a way of tracing these RC data throughout the equipment's life, from installation to retirement.

Focus on Easy, Low-cost Maintenance and Operation

Although MP design aimed at keeping maintenance costs down tends to conflict with initial cost reduction design, keep the following key points in mind:

- The design must make deterioration and other defects easy to spot (design emphasizing high accessibility).

	Function		
	1. Positioning of parts, moving parts to setup position	2. Positioning of contact points, layout of contact points, angle of contact points	3. Welding Applied pressure: Pressing force Reverse-force backup Heat input: Resistance welding method Resistance welding amount
	Conventional Method		
	Cart method: Position on cart, then do block replacement	Check with fixed-position gun	Air cylinder, frame structure, series welding
Replacement			
Can it be moved to another process?			
Can it be set up for flow processing?			
Can it be modified for the use of general purpose parts?		○ Check if robot can do it; examine gun arrangement mechanism	
Can it be done using a simple automated device?			
Can other means be used for some parts?			

Simplification			
Can it be eliminated?	○		
Can the product structure be changed?	(bracket continues)		
Can the number of contact points be reduced?	Improve cart structure to facilitate positioning; see if block can be set to the direct weld position		
Can the number of guns be reduced?	(bracket continues)	○	○
Can the number of spindles be reduced?	(bracket continues)		Check into direct welding using portable gun
Can the contact point precision requirements be loosened?	(bracket continues)		(bracket continues)
Can the welding method be changed?	(bracket continues)		○
Can the number of functions be reduced?	○		
Can the functions be divided differently?			
Combination			
Can independent right and left functions be combined?			
Can the jigs be made more uniform?			
Can it be combined with another process?			
Can it be combined with another part?			
Can the number of contact points be reduced?			
Can other functions be taken over by robots?			
Can the operator/machine combination be improved?			

Figure 4-35. Issues for Reducing Auto Body Equipment Costs

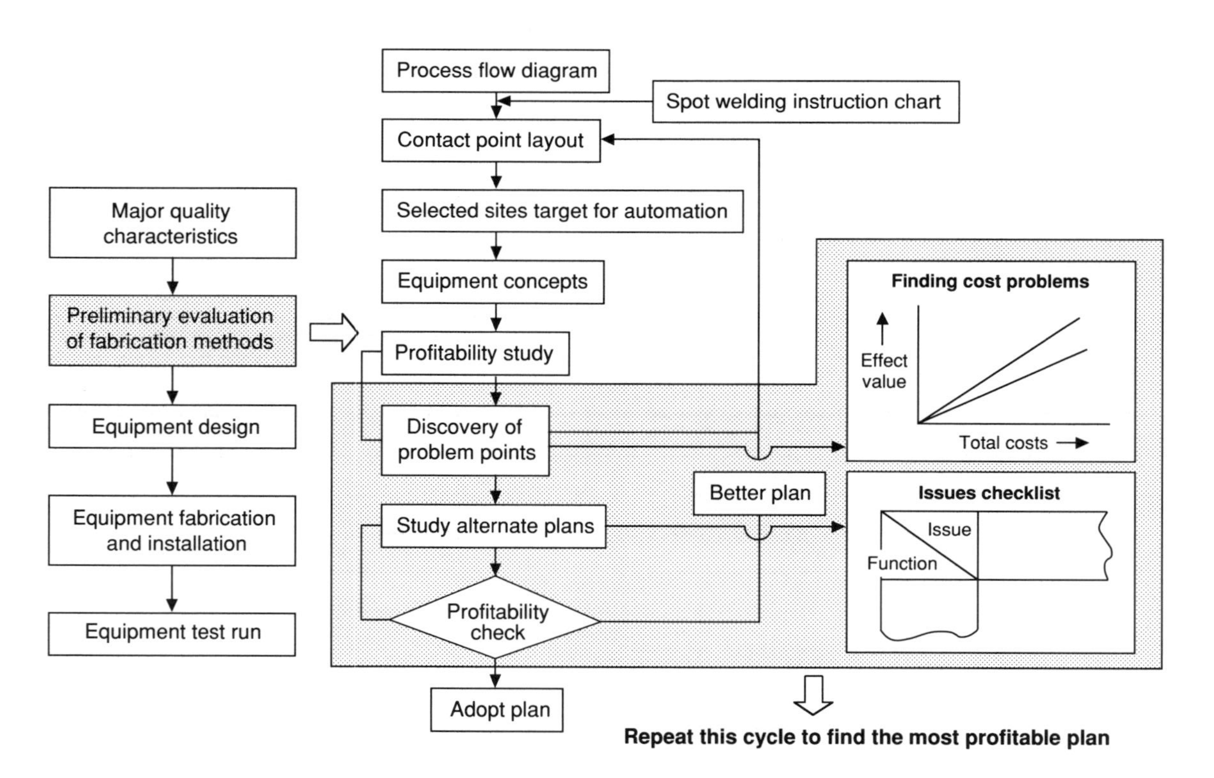

Figure 4-36. Flowchart of Improvement Plan Review Process

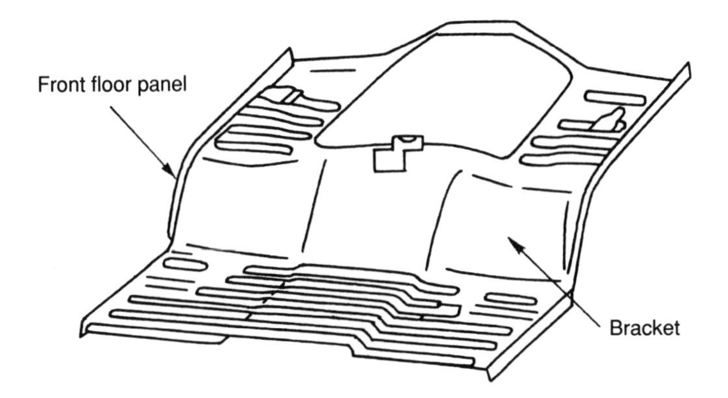

Figure 4-37. Front Floor Process

- Custom (consumable) parts should be designed for easy replacement. (Even ordinary parts can sometimes be included in the stock of spare parts to enable RC reduction, if they are designed for block replacement or unit replacement).
- The design should make processes easy to modify (minimum cost for modifying processes, with minimum preventive maintenance and personnel changes).
- The design should limit the possibility of accelerated deterioration by combining low cost and high durability.
- The design should eliminate stress as much as possible, such as in sputter, wiring, or water leakage, through use of limiters (detectors), by facilitating lubrication (or eliminating the need for lubrication), and so on.
- The design should be safety-oriented.
- The design should facilitate operation and reduce the potential for operator errors.

In sum, the design should promote human-machine system goals.

Function			Equipment concepts
1. Positioning of parts • Carry to setup position • Use positioning mechanism	**2. Positioning of contact points** • Layout of contact points • Angle of contact points	**3. Welding** • Applied pressure —Pressing force —Reverse-force backup • Heat input —Resistance welding method —Resistance welding amount	
Conventional method			
Cart method: Position on cart, then do block replacement	Check with fixed-position gun	Air cylinder, frame structure, series welding	Multi 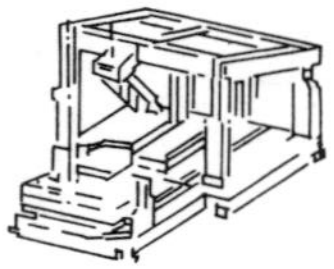
Issues			
• Movement of heavy objects • Use of large equipment	• Need separate base for attaching each gun • Need large frame for attaching guns	• Large structure • Line up several electrodes at spot weld points	Two simple robots plus positioning jig 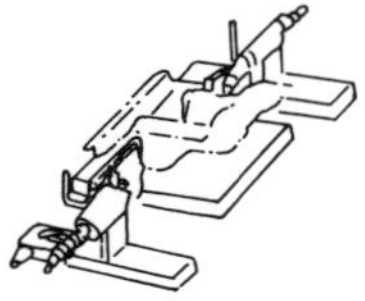
Improvement plan I			
Use fixed method	Install branching device to each gun	Install gun on main unit	

Issues Need two robots	Numerous obstacles	Large gun	One commercially advanced robot plus positioning (backup) jig
Improvement plan II Work from the top side of the product ➡	Air cylinder: use lightweight gun with one air cylinder	Frame structure designed to support one gun	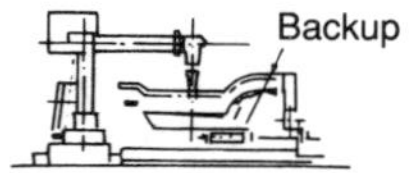 Backup
Issues	Requires sophisticated, articulated robot		Gun positioning frame plus positioning (backup) jig
Improvement plan III	Gun moved by hand ⬇ Gun position determined by backup stand		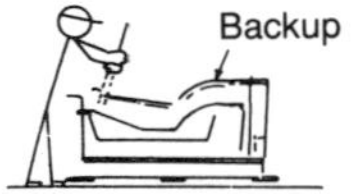 Backup
Improvement plan IV	Replace with low-cost robot ⬇ • Insufficient functions • Add supplementary functions	• Uniform contact point angle (corresponding to product design)	One simple robot with supplemental function plus positioning (backup) jig 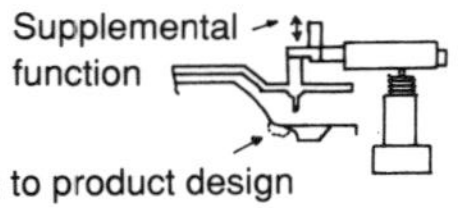 Supplemental function Corresponds to product design

Figure 4-38. Process of Selecting Improvement Plans

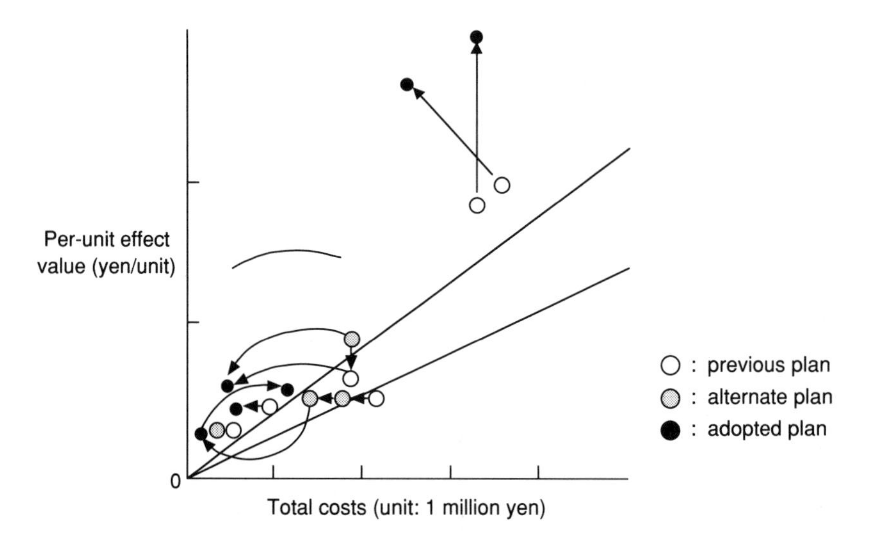

Figure 4-39. Results of Examining Spot Welding Sites

LCC Design Priorities

Obviously, every aspect of LCC design described above cannot be implemented for all equipment. To sort out the design priorities, consider which aspects of LCC design are needed most for which pieces of equipment and evaluate the equipment to be designed (beginning at the cost planning stage). Establishing design priorities is an essential part of creating a design evaluation system.

Economic Analysis

Although there are various ways of evaluating investment economy, the engineering economics approach is the best method from the LCC design perspective. If you adopt a policy that engineering economics will be applied to evaluate planned investments, then all equipment design engineers should become proficient in the discipline.*

Review of Design Standards

Almost every company has its own set of design standards. In adopting the LCC design approach, a company must evaluate the process by which their standards were created and consider how they relate to technologies such as computer-aided design. Design standards are essential, but if taken too seriously they can hinder innovative design work. When engineers are committed to fresh thinking and the LCC design approach, they can generate new design standards that combine high reliability with low costs.

---

* A thorough explanation of engineering economics is beyond the scope of this book. For more information consult the works of Shizuo Senju, chief proponent of engineering economics in Japan. Two of his books are available in English from Quality Resources: *Economic Engineering for Executives*, co-authored by Zentaro Nakamura (1990), and *Profitability Analysis: Japanese Approach*, co-authored by Tamio Fushimi and Seiichi Fujuta (1989).

5

Design for Flexibility

As noted in Chapter 1, today's companies face a harsh and quickly changing business environment, and these challenging trends are expected to become even more pronounced. Consequently, a key factor in corporate survival is the ability to respond flexibly to environmental changes. Equipment design must be flexible with regard to three types of change:

- Change in production volume (yield)
- Changes due to product diversification
- Changes due to successive models

Case studies reflecting each of these types of change are included as part of the chapter discussion. Each case study focuses on the design approach, with the assumption that the design mission (the target values for cost, quality, production yield, and so on) has already been determined by management.

DESIGNING FLEXIBILITY FOR VARIABLE YIELD

Generally, a new equipment design is expected to fulfill a specified production capacity (or planned output). In some cases the rate of production, averaged over the product life cycle, is roughly equal to the constant figure estimated at the production planning stage. In most cases, however, market changes and other environmental forces cause production yield levels to fluctuate widely from month to month or year to year throughout the product life cycle. Companies with equipment too inflexible to respond to these fluctuations suffer major losses or invest hastily and unwisely in additional equipment. (Case study 5-1 reflects the latter situation.)

Losses Incurred as a Result of Variable Yield

Companies typically determine equipment capacity to meet a constant production yield figure established as the planned yield. Figure 5-1 illustrates what happened in one company when a product model change was followed by about two and a half years of production boosts well beyond the planned yield. The company responded by adding an overtime production shift, investing in more equipment to boost production capacity, and hiring more workers. After this two-and-a-half-year boom period, however, they experienced a sudden drop in the required yield. As a result, the company suffered various extra costs due to yield fluctuation. When demand was low, for instance, the company had to pay extra processing labor hours per product unit (Figure 5-2) and higher per-unit energy costs (Figure 5-3).

The problems arising from variable yield can be summarized as follows:

1. Investment loss due to variable yield (Figure 5-1)
   - Redundant investment in additional equipment
   - Reduced equipment efficiency during periods of lower production yield
2. Processing labor hour loss due to variable yield (Figure 5-2)
3. Energy-related loss due to variable yield (Figure 5-3)

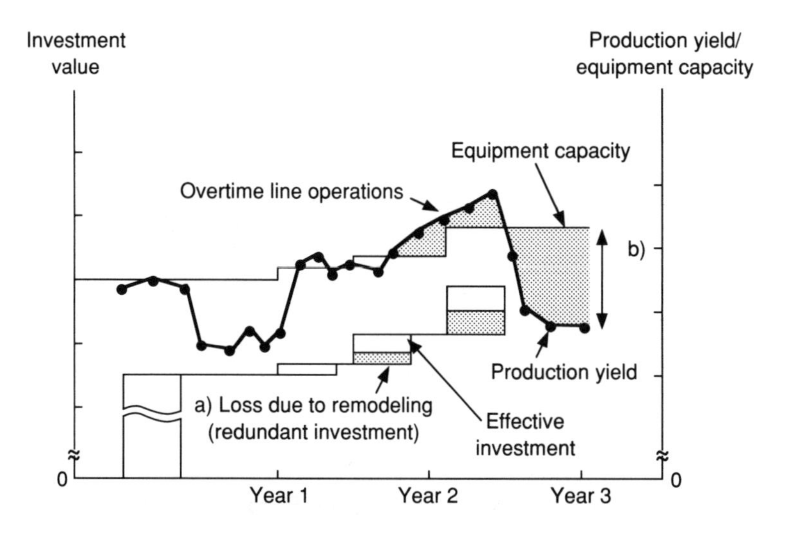

Figure 5-1. Production Yield and Capacity Response for Current Product Model

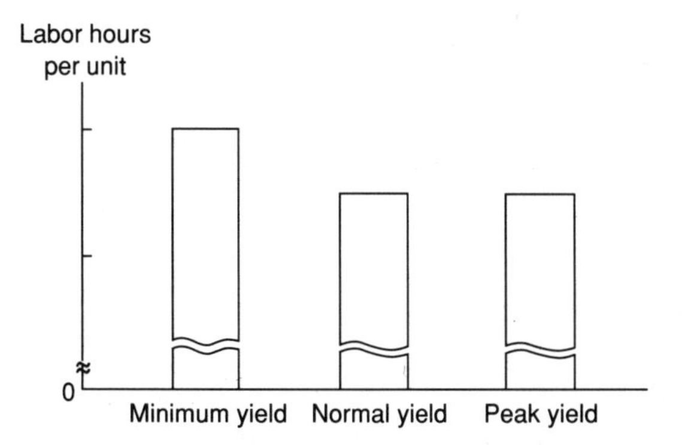

Figure 5-2. Processing Labor Hours per Unit (Current Product Model)

To produce only what is necessary, when it is needed, in the required amount (i.e., just in time), without having to purchase additional equipment when product demand exceeds equipment capacity, you must consider flexibility and expandability from

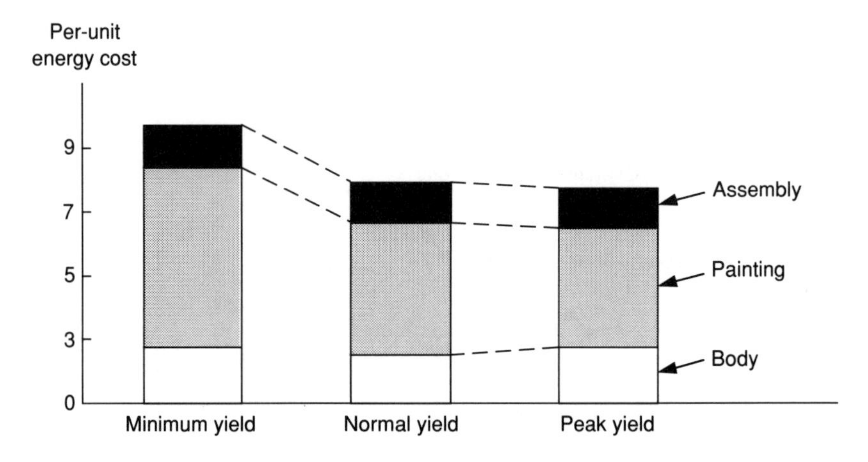

Figure 5-3. Per-unit Energy Costs (Current Product Model)

the beginning of the equipment design process. At the same time, be careful not to overinvest in equipment, because that will lower the return on investment whenever the production yield drops significantly. Equipment design must also allow for the possibility of efficiency-related loss due to overcapacity and overinvestment. Thus, flexible design for variable yield means equipment that is economical in terms of initial investment and operating costs and that can also adapt to variance in yields.

Design Approach for Variable Yield

The following approach can be adopted by a company building a new production line to turn out a new product (a model change that is fairly similar to previous models).

Step 1: Estimate Production Yield

First draw up a time-series demand trend pattern, based on data on previous similar products, market surveys, and other information (Figure 5-4). Then:

- Estimate the maximum and minimum production yields throughout the product life cycle.
- Estimate the range of production yield variance under ordinary conditions.

If it is too difficult to make precise predictions but you have a rough idea of the maximum and minimum production yield levels, try the "design using SA" approach described in Chapter 4 as an alternative method.

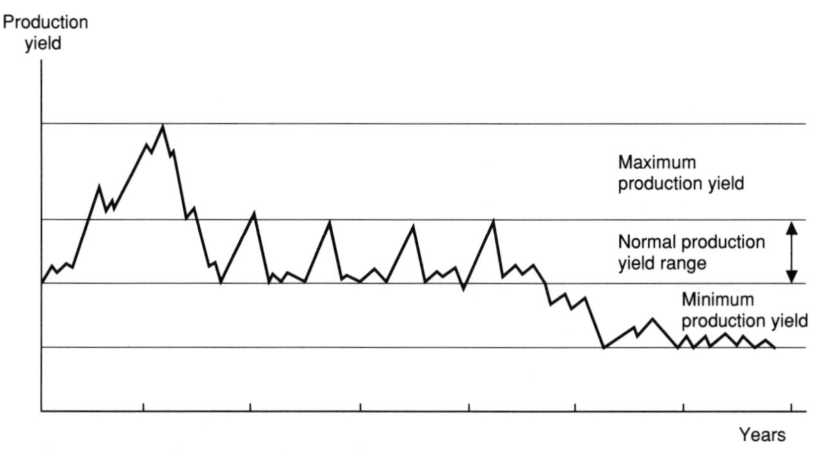

Figure 5-4. Production Yield Range

Step 2: Analyze Loss Due to Variable Yield

Using data gathered from past experiences and results from similar products, identify the kinds of loss that occur as a result of variable yield.

Analysis of investment loss. This includes analysis of results and effectiveness of additional investments made in response to higher production yields. It also includes analysis of investment loss (wasted investment) caused by lower production yields.

Analysis of operating loss. This includes various losses that occur when fewer product units are produced, such as equipment operating loss, labor hour loss, material loss, and energy-related loss.

Step 3: Establish the Initial Production Capacity

Decide how the initial production capacity (when production of the new product begins) fits into the yield variance pattern that runs through the entire product life cycle. To do this, study the analyses of the estimated production yield pattern and the loss due to variable yield. Then establish the most economical production yield point as the provisional initial production capacity. (The most economical yield point is the point at which the sum of additional investment to boost yield and loss from reduced yield is lowest.)

Step 4: Estimate Loss Due to Process Design and Variable Yield

Using the designs for processes and equipment that were based on the provisional initial production capacity, look at the measures taken in the past to adapt those processes and equipment when the production yield fluctuates. Then, refer back to your step 2 analysis of loss due to variable yield and predict what type of loss is likely to occur at which processes. At the same time, consider optimal ways of responding to this estimated loss. What kind of flexibility is needed to adapt to what kind of factors? For instance, simply adding or subtracting a few robots may provide the needed flexibility.

Step 5: Work out the Details of Responding to Variable Yield Loss

The analysis in step 4 should help the design team to select from among many different responses to variable yield loss based on different product or process characteristics and to develop a general response. With this basic orientation established, the team can study the feasibility of flexible fabrication methods, equipment, layout plans, and other design elements that are consistent with that orientation. It is essential to limit initial investment to the amount required to establish the estimated initial production capacity.

At the same time, ensure that production boosts can be handled with minimal additional investment, and that other costs and production dips can also be accommodated with only minimal increases in per-unit operating loss. Case study 5-1 illustrates how a production yield pattern was used to work out a response to variable yield.

Step 6: Review Progress

The responses chosen in step 5 may lead to changes in the estimated required level of initial production capacity. In addition, the range of responses to choose from may change if the estimated values for production yield variance prove unreliable. If these production yield variance estimates remain vague, apply the method for design using SA described in Chapter 4 to analyze the safety and economy of your response measures and to determine final, concrete responses.

Case Study 5-1: Design for Variable Yield

In this example (illustrated in Figure 5-5), new product planners set up different levels of planned production yield (minimum, normal, and peak) on the basis of results from previous products. Working from this production yield pattern, they provided for the following approaches in process design, equipment planning, and layout planning to reduce variable yield loss:

- Use human labor to respond during temporary peak yields.
- Combine human and machine labor in different ways to accommodate small month-to-month variances.
- Transfer staff or remove some machines in response to reduced yield. Control energy consumption in accordance with the number of machines currently in use.

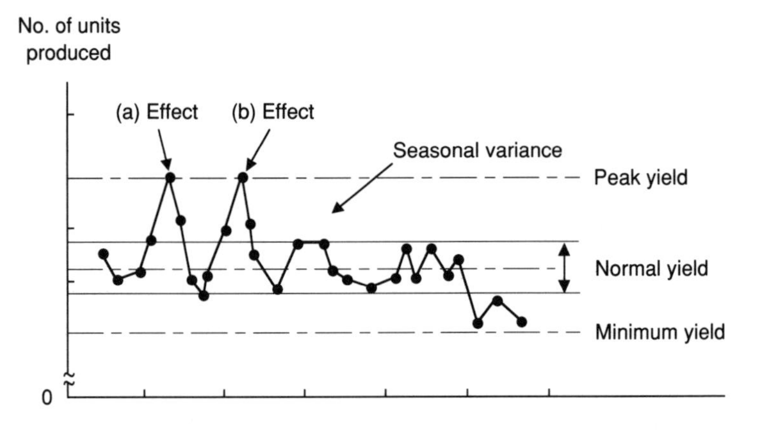

Figure 5-5. Planned Production Yield for New Automobile

The specific conditions under which these measures were implemented are described below.

Process Design for Flexible Response to Variable Yield

The company implemented the following process design approach to avoid purchasing additional equipment to increase production capacity (Figure 5-6).

Empty-room method. Manual processes and automatic processes were designed to overlap to a certain extent, so that fluctuations in production volume could be met by adding or subtracting workers. This method costs far less than adding new processing equipment to the line during production boosts.

Designing expandability into production lines. Anticipating that production yield may grow to the point where new processing equipment must be added, the team designed the connection between main assembly and subassembly lines so that the conveyance route could be changed easily to accommodate additional equipment.

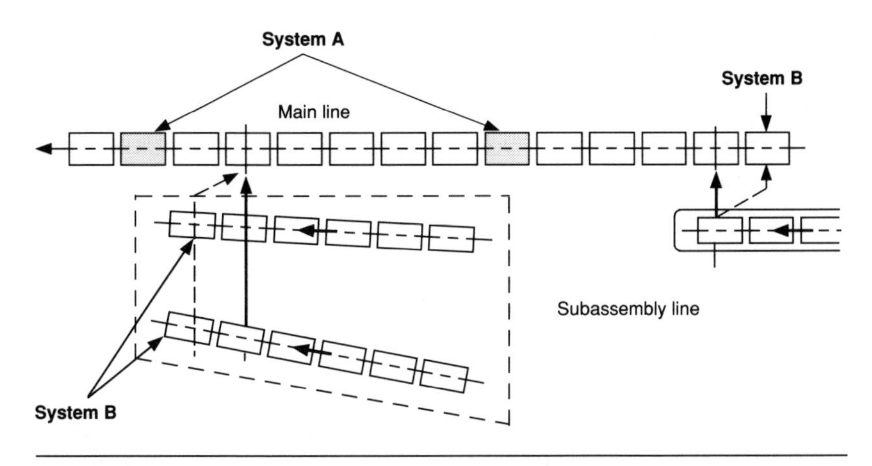

System A:
Processes that take over surplus operations from the robots

System B:
Additional processes (handled by extra staff at peak periods)

Figure 5-6. Process Layout Plan for New Automobile Bodies

Automated Line Designed for Flexible Work Load

Previously, the company found that when production volume dropped, some of its machines became idle, and the overall equipment capacity utilization rate also dropped. On a robot-equipped line, for example, it took time to reteach the robots whenever the production schedule was changed to adjust the yield, so production schedule changes could not be made promptly. Even when they tried to substitute human labor for certain specialized, automated machines, equipment remodeling costs were still high.

To solve these problems, they expanded their use of general-purpose automated equipment and switched to programming that could be redistributed easily among different machines. These steps enable prompt work load changes to accommodate fluctuations in yield. Furthermore, if the required yield takes an unusually large downturn, the company is now able to perform rapid changeover procedures to remove some equipment from the line. The company also upgraded its robots by purchasing or developing robots that can either handle two welding guns at once or easily exchange welding guns. This helped maintain a high-capacity utilization rate. Figure 5-7 shows an example of improved capacity utilization on a robot line.

Coordinating Equipment Operation Conditions with Production Yield

The painting process requires large amounts of energy. Normally, painting quality is determined by painting time, which is determined by processing speed within the booth multiplied by booth length. When production yield is increased, processing speed is accelerated and booth length is increased. When production yield drops, booth length is shortened but energy costs remain the same, even though fewer units are being produced. In other words, per-unit energy costs go up.

The company responded to decreased yield in this case by slowing down and shortening the operation zone. This action helped minimize per-unit energy consumption loss. The increased cost of dividing the operations into small segments was still low compared to running costs. For example, when the company had to lower its operation rate to 70 percent of full

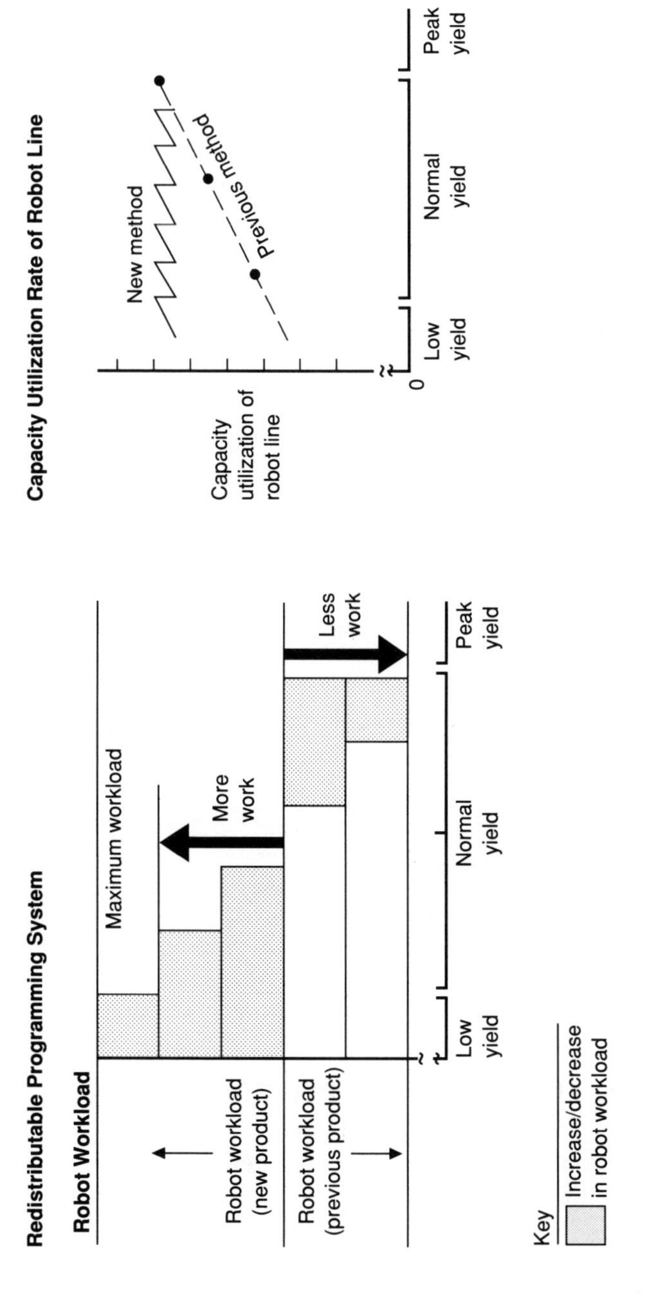

Figure 5-7. Improved Capacity Utilization Rate on Robot Line

capacity, it was also able to lower its energy costs about 30 percent (see Figure 5-8).

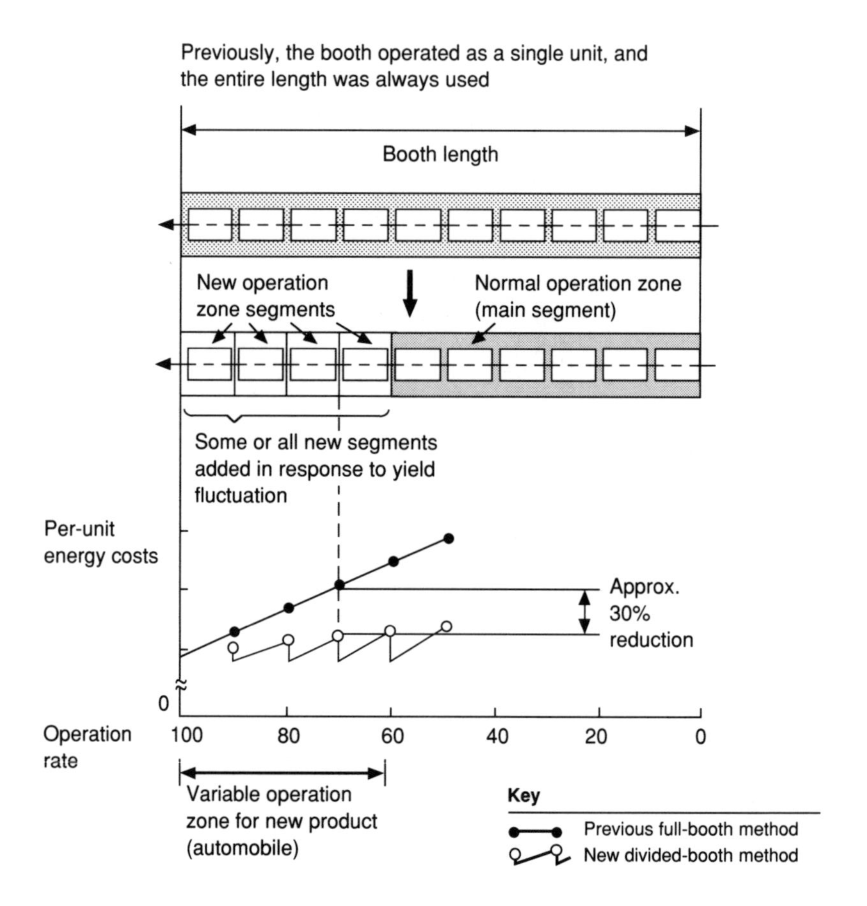

Figure 5-8. Painting Booth Operation Zone

DESIGNING FLEXIBILITY FOR DIVERSE MODELS

With today's diversified market demands, companies that want to remain competitive must respond with more diverse products. Companies have found, however, that installing specialized equipment or specialized production lines for new products require too much plant investment to maintain profitability. Moreover, another recent market phenomenon — shorter product lives — makes it even harder to obtain a profitable return on the capital used to turn out the products.

This situation created the need for mixed production systems, that is production equipment and production lines that can turn out more than one kind of product. Mixed production systems do not work well, however, if the changeover between product models takes too long, if the capacity utilization rate drops too low, or if changeover-related labor hours and increased inventory add too much to the total production cost. Mixed production systems are also susceptible to the following problems:

- Proliferation of parts variety can greatly increase wasted labor hours if workers must walk farther to pick up and return parts. In addition, greater variety in specifications creates gaps in assembly labor time between processes, which also contributes to idle time.
- In the painting process, for example, standards set to accommodate the product models that are the most difficult to paint boost material and energy costs for models that can be painted easily under more economical standards.

A highly automated production system that can also handle variety-related problems is known as a flexible manufacturing system (FMS). When equipment design is oriented toward mixed production with minimal operating loss in terms of labor hours, materials, and other cost factors, it is referred to as design for flexibility.

Example of Loss Related to Mixed Production

A certain factory's products consist mainly of automobile shock absorbers and suspension struts. The factory has had to deal with major product changes, however, as the number of different product models grew to 2,370, and as major specification revisions called for more and more electronic components in the products.

At first, the factory tried to maintain its basic job-shop, batch-processing production system by adding new processing for electronic parts, giving some work to other factories, and shifting the production flow. As shown in Figure 5-9, however, this greater complexity led to a corresponding doubling or tripling of costly conditions such as more stock points, longer

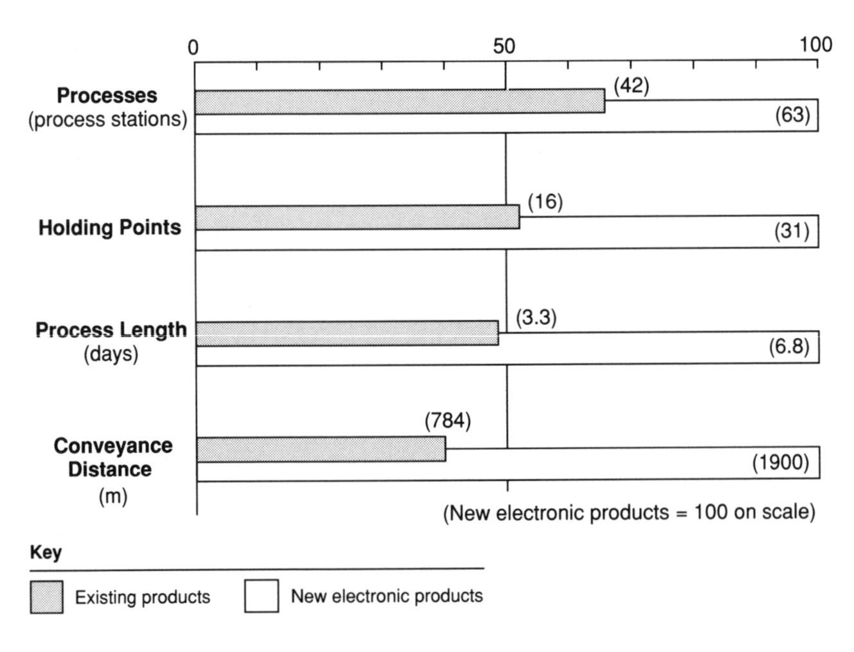

Figure 5-9. Process Analyses for Existing Products and New Electronic Products

processing periods, and longer conveyance distances. This produced a negative overall effect on the production flow.

In addition, the shift toward more electronic components led to a sudden 132 percent rise in work-in-process inventory, as shown in Figure 5-10. This alone contributed greatly to the increase in the numbers of processes and points and to other factors that adversely affected the production flow.

As for product quality, predelivery defect analyses showed that these flow-related problems contributed at least indirectly to 45.1 percent of all product defects found prior to delivery. Moreover, 20.3 percent of these defects were found to be directly caused by poor production flow conditions (see Figure 5-11). Figure 5-12 shows that 58.6 percent of the defects discovered within the processes where they occurred were attributable to the poor flow conditions. Out of the total defects, 21.8 percent were still caused by flow-related problems such as excessive in-process inventory, conveyance, and packing type.

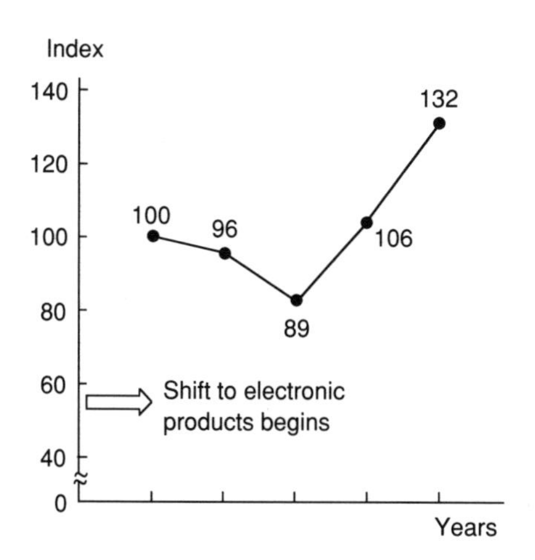

Figure 5-10. Effect of Shift to Electronic Products on Work-in-Process Inventory

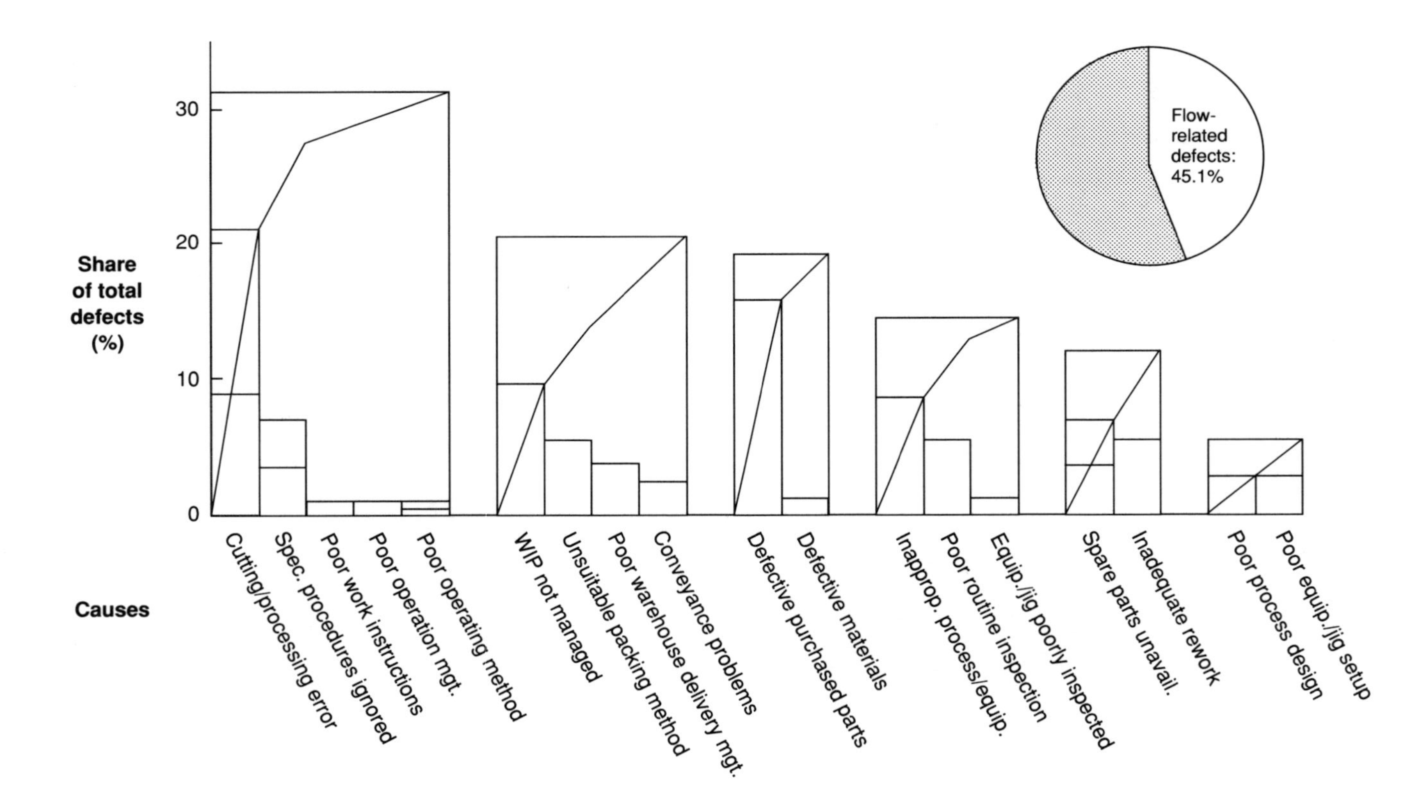

Figure 5-11. Analysis of Pre-delivery Defects

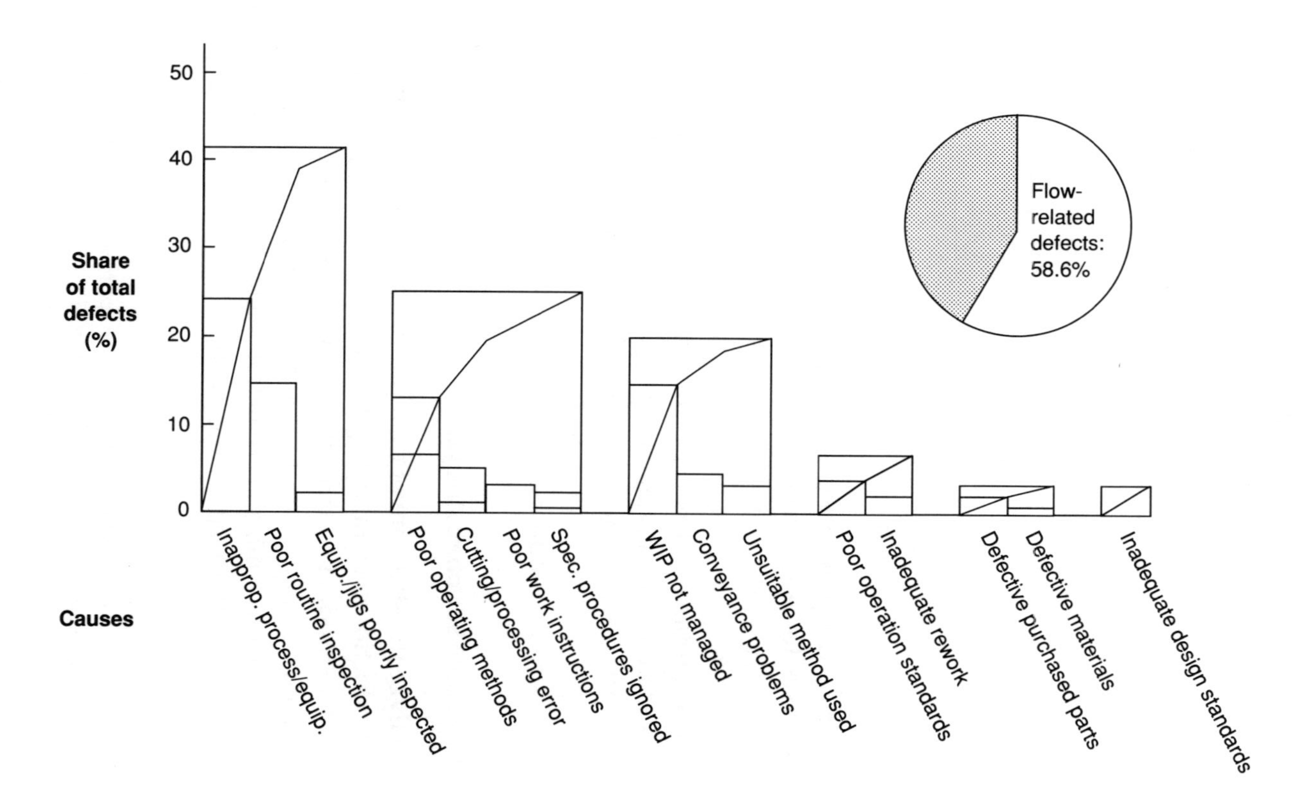

Figure 5-12. Analysis of In-process Defects

Basic Approach to Mixed Production

This approach is premised on the factory's existing job-shop, batch-processing manufacturing system, but it can also be applied in factories that have established continuous-flow production from the first to last processes and factories that are in the process of switching to a one-piece flow. A step-by-step description follows.

Step 1: Process and Flow Analyses for Each Product's Conditions

As in as the previous example, first carry out process and flow analyses for each product to identify current problems — namely, loss related to product diversification — and to pinpoint obstacles to mixed-flow production strategies that might solve those problems. It is especially important to analyze factors such as processes, cycle time, flow, conveyance methods, in-process inventory, operation methods, and equipment performance, and to obtain clear, quantitative data on loss caused by current, product-specific abnormalities.

Step 2: Product Structure Analysis

Analyze the structure of each product to be included in the mixed production system, trying to find ways to unify, simplify, or abbreviate product specifications so processes can be reduced or combined. In short, aim to make each product's structure more conducive to a mixed production system. The engineering approach used in this step is called variety reduction.*

* Editor's note: for a detailed discussion of this approach see Suzue, Toshio and Akira Kohdate, *Variety Reduction Program (VRP): A Production Strategy for Product Diversification* (Cambridge, Mass., Productivity Press, 1990).

Step 3: Product Grouping and Process Design

Based on what you have learned so far about the products' structure and about similarities among processes, group together product models in whatever way is most conducive to mixed production. Once products are grouped together, you can proceed with a concrete process design to establish mixed production. The engineering method for this is called group technology, or GT.

In grouping products and designing the processes, distinguish the less practical proposals to get a clearer idea of which products within the group and which processes are likely to pose obstacles. From the perspective of production yield or other factors, determine the efficiency of including one or more specialized lines for certain products.

Step 4: Line Composition

Make a close and careful study of possible line compositions for each group or family of products to find which is best. If necessary, you may want to work within product groups to estimate the production yield, takt time balance, personnel, and space requirements of individual product models. Keeping in mind the technical hurdles, feasibility problems, and other restrictions identified in step 3, determine whether it is best to have one line or several, what kind of flow pattern to build, and other composition-related factors.

Step 5: Clarification of Process-specific Flexibility Conditions

With an analysis of technical hurdles from step 3 and a general idea of the line composition from step 4, you are ready to examine and clarify the demand conditions for each process within each line. Since several parts or products pass through some of the same processes, the line must be designed with flexibility to process or assemble all the intended products.

In mixed production lines, for example, there are many minor differences in how each product is handled at each process. Problems that typically arise include large gaps between different products' processing times (or processing precision tolerances), different product-specific processing procedures within the same process, or different changeover procedures or operation standards that require a change of blades or drill bits. Consequently, you need to clarify and organize the conditions needed to ensure flexibility to handle the full range of target products. These conditions include fabrication methods, equipment performance characteristics, jigs and tools, measurement methods, changeover procedures, and conveyance and workpiece positioning methods.

In the remaining three steps, you will work out the details of these process-specific flexibility conditions and estimate the corresponding costs (cost targets).

Step 6: Layout Design

Using the line composition outline from step 4, work out a specific layout design that meets the flexibility requirements identified in step 5. The design should also take into account restrictions due to factory layout and the locations of utility, plating and painting, drain water processing equipment, and any other equipment that cannot be moved easily.

While examining the flow of goods within processes, also consider key factors that affect the layout design, such as material handling between lines and processes, and methods of supplying, storing, and shipping materials, procured parts, and finished items.

Step 7: Design of Control and Management Systems

A system design must be created that satisfies the flexibility requirements from step 5 at each level, from process-specific control to the entire production management system. Establish control methods that make for efficient management and decide how much computerization is warranted. For example, consider how the control and management systems will incorporate different control levels and methods, scheduling, production orders, progress evaluations, corrective measures, and the like.

Step 8: Study of Fabrication Methods and Detailed Design

Now you are ready to resolve the technical issues that hinder implementation of the plans developed in steps 6 and 7, while satisfying all the flexibility requirements set forth in step 5. Start by studying and selecting fabrication methods, then address the following issues:

- Equipment specifications
- Jigs and tools
- Automatic conveyance devices
- Changeover work
- Process supervision, abnormality response methods, and so on

Two case studies of design for flexibility approaches are provided in this section. Like the earlier example of loss due to product diversification. Case study 5-2 is concerned with eliminating loss that occurs in a job-shop factory that must handle a wide variety of products. The factory opted to establish mixed production using one integrated line and the one-piece flow method. The factory faithfully followed the basic approach described above. This case study has been simplified somewhat to provide a clearer explanation. Case study 5-3 is another application of the basic approach just described. In this example, a factory that had already established an FMS line used a mixed-production approach to add new products to the line's production flow.

Case Study 5-2: Mixed Production

A company's shock absorber plant consists of two factories: one that produces 650,000 lightweight-vehicle shock absorbers per month (in 1,940 models), and one that turns out 350,000 suspension struts per month (in 430 models). Figure 5-13 is a flowchart of the strut manufacturing processes.

Within two and a half years of introducing TPM, these factories had dramatically reduced the six big losses and were able to maintain high reliability in their equipment. In addition, like the factory cited on page 258, these factories managed to eliminate loss inherent in their conventional job-shop, batch-production manufacturing system by implementing a long-planned shift to mixed production (using a single integrated one-piece flow line from first to last process). Figure 5-14 outlines the development program for the shift in these factories toward TPM and mixed, one-piece flow production.

Analyses of Original Processes and Flow

Figures 5-15 and 5-16 show the results of the original process and flow analyses. The production system layout at the two factories was complex. Figure 5-16 indicates that the layouts included large inventory stockpiles between and within the processes. The improvement team identified the following problems and mixed production development themes based on the various process and flow analyses:

- Poor takt balance — acceleration is required.
- Changeover time is too long and is done at too many processes — develop single-minute, or one-touch changeover techniques.
- Processes are separated like small islands — connect processes directly.

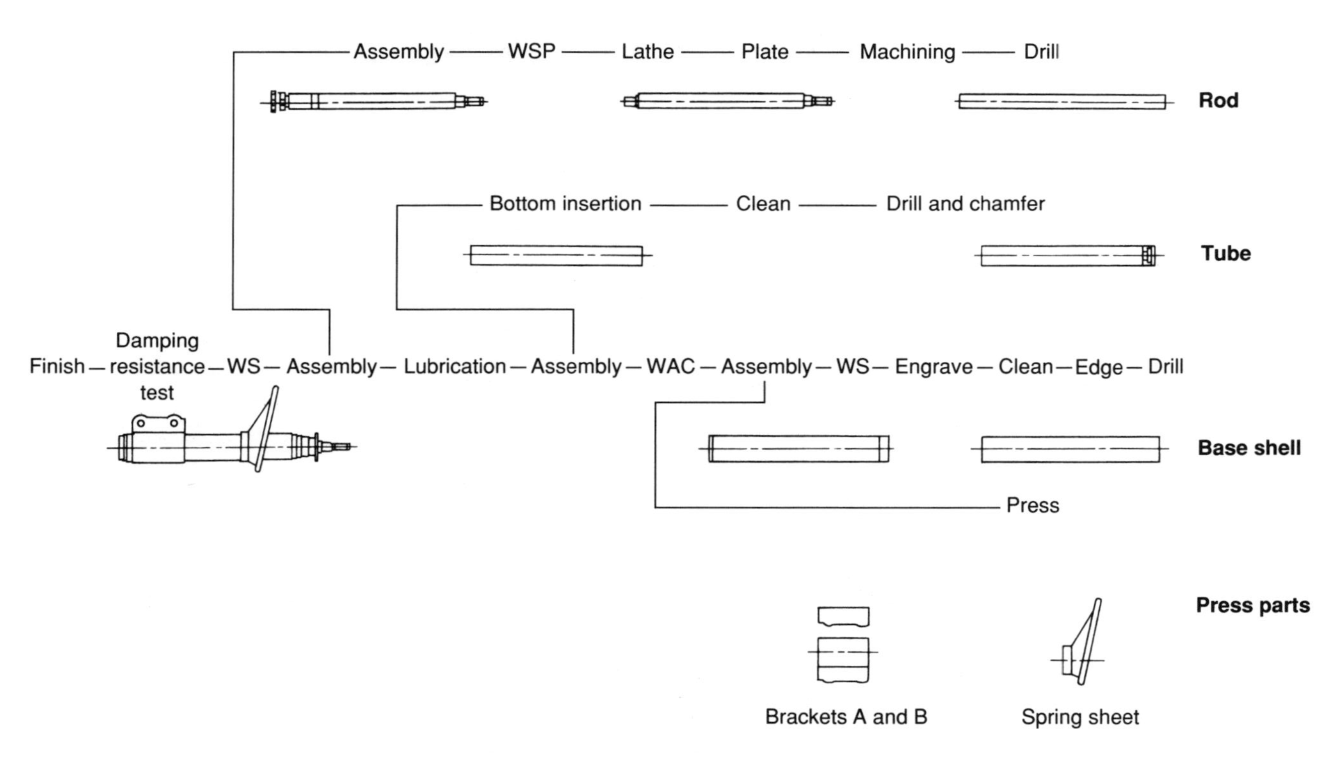

Figure 5-13. Flowchart of Strut Manufacturing Processes

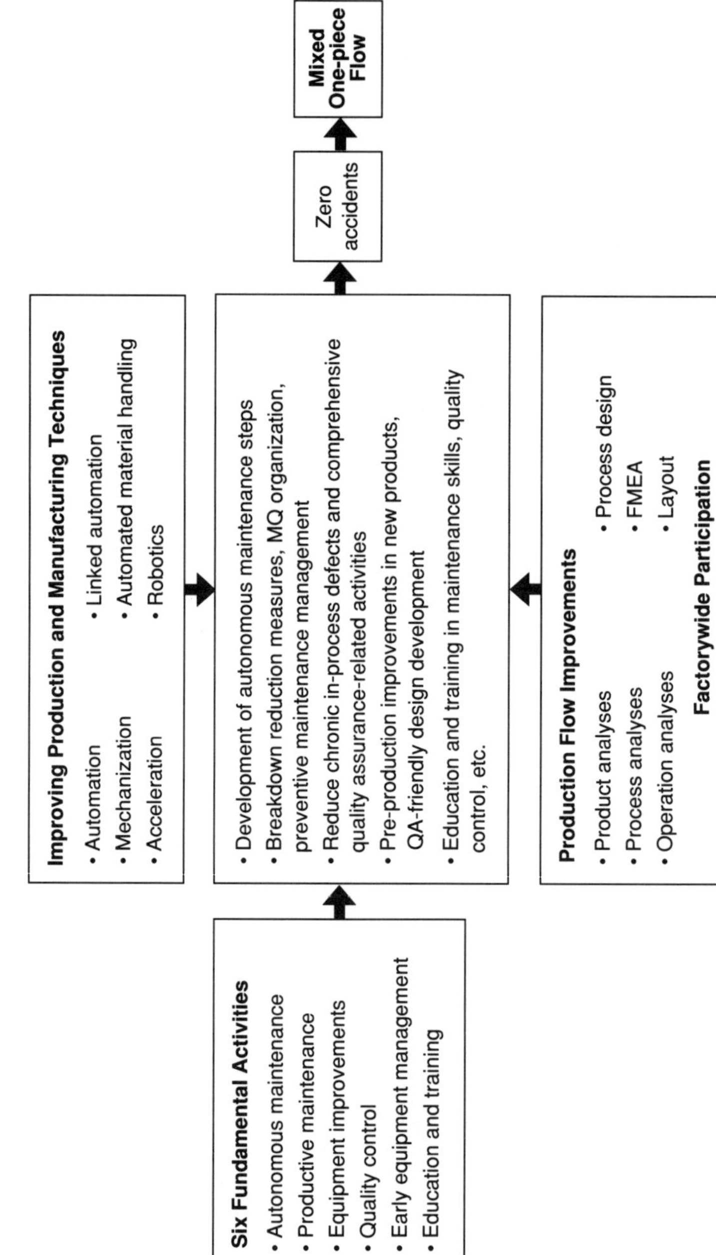

Figure 5-14. Shift Toward TPM and Mixed, One-piece Flow Production

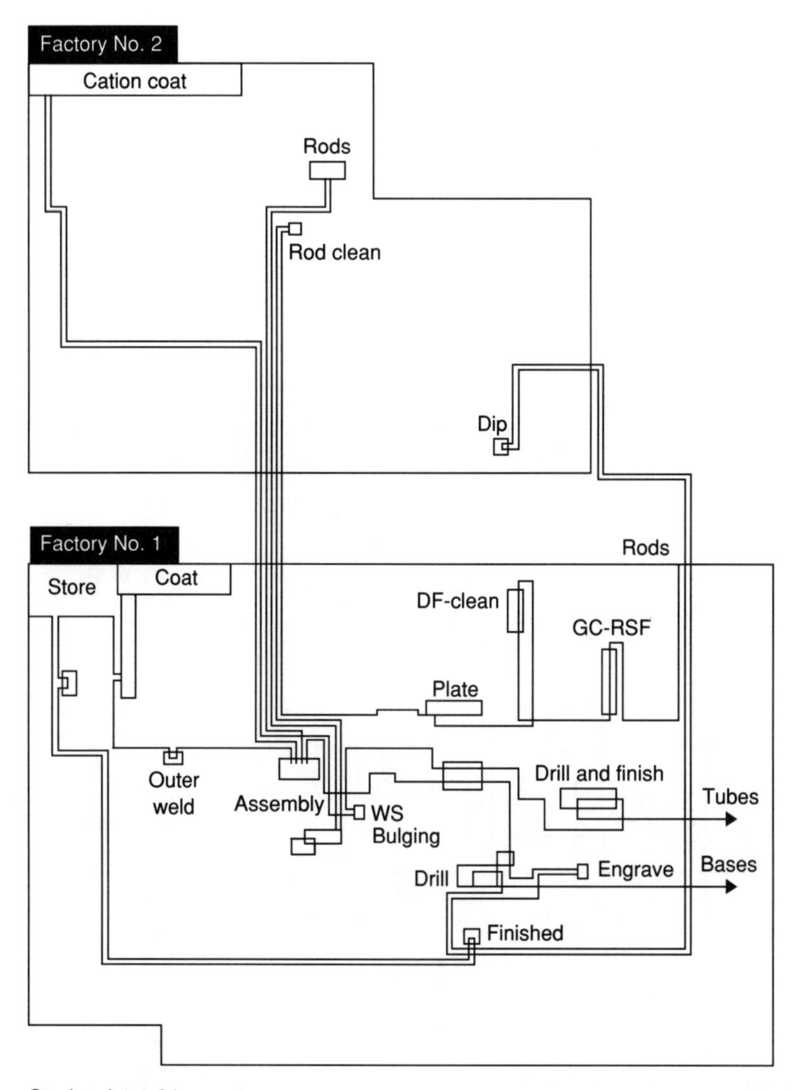

Stock points: 31
In-process inventory: 5,000 units
Total process length: 6.8 days
Total conveyance distance: 1,900 m

Figure 5-15. Process and Flow Analysis for Model Product

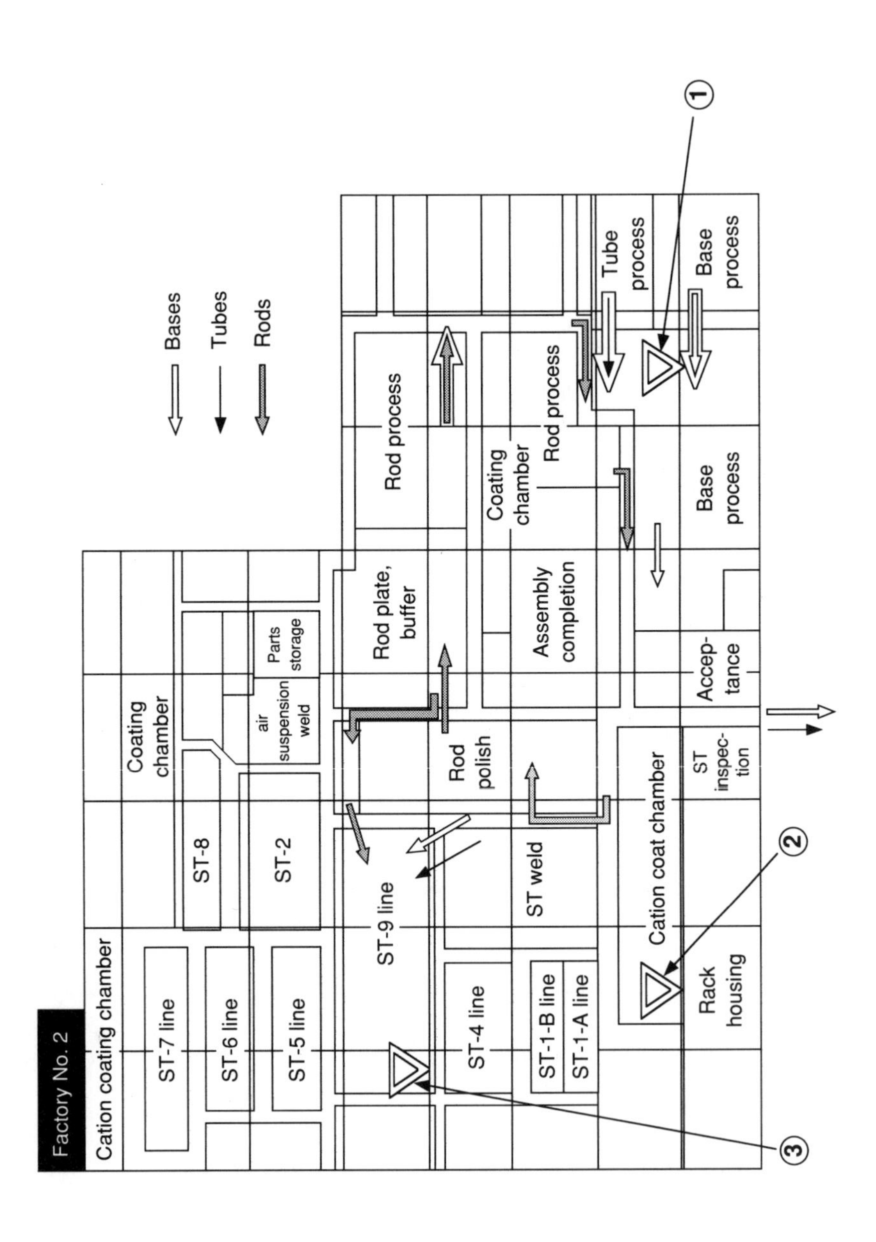

Bases
Tubes
Rods
Factory No. 2
Cation coating chamber
ST-7 line
ST-6 line
ST-5 line
ST-9 line
ST-4 line
ST-1-B line
ST-1-A line
Rack housing
Cation coat chamber
ST weld
ST-8
ST-2
Coating chamber
air suspension weld
Parts storage
Rod polish
Rod plate, buffer
Assembly completion
ST inspec-tion
Accep-tance
Coating chamber
Rod process
Rod process
Base process
Tube process
Base process
①
②
③

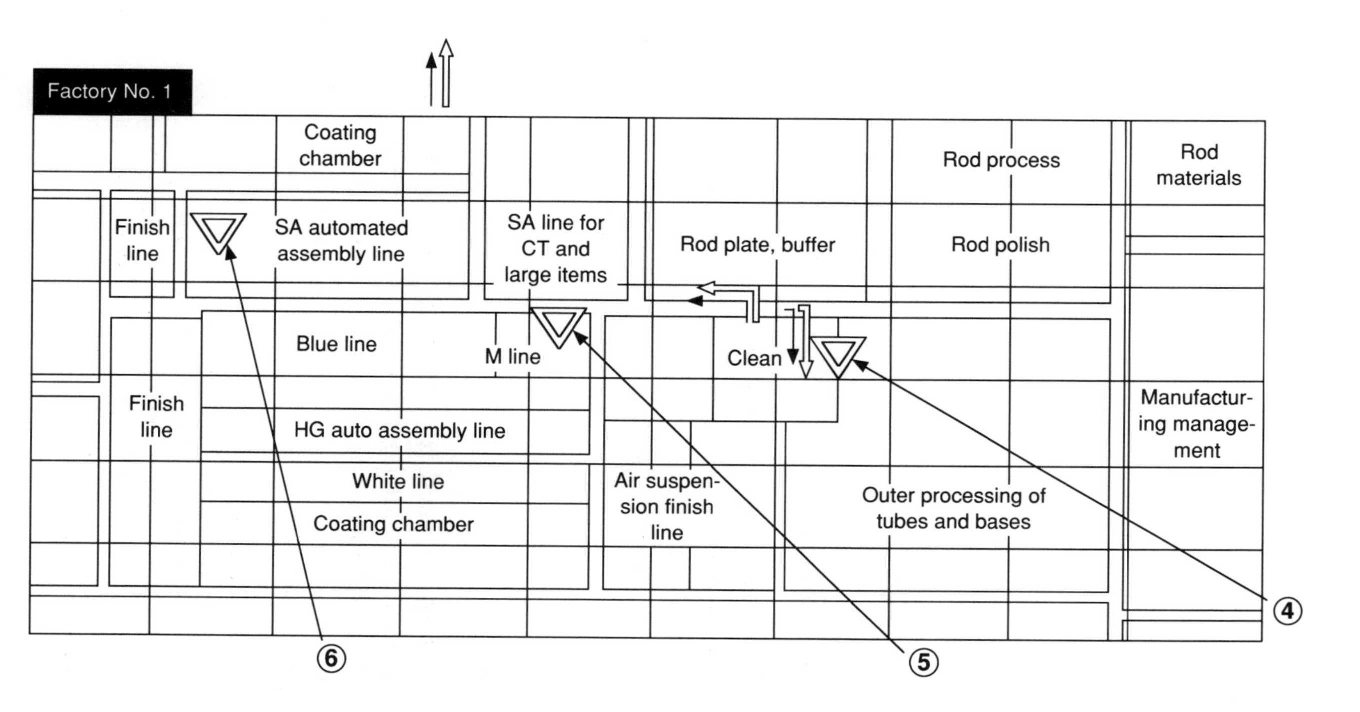

Figure 5-16. Current Layout

- Some processes have numerous defects — implement poka-yoke* devices and defect prevention measures.
- Too much inventory between processes — improve production flow by using one-piece flow production.

Line Organization

Next, the team used GT and the above analyses to group the products and reorganize the line around new process designs. The results of this reorganization are shown in Table 5-1, which lists the product lines connected to one-piece flows, mixed lines, or special synchronized lines.

Clarification of Process-specific Flexibility Conditions

At this point, the team reorganized the list of technical obstacles to mixed production in terms of specific needs of the line organization just described. Table 5-2 lists examples of how these were addressed systematically through improvements implemented on the shop floor or with the assistance of technical staff.

Layout Design

The company used its new line plans as the basis for designing a new layout for the entire floors of both factories. Several thousand pieces of equipment were shifted. Figure 5-17 shows the new layout. Compare this with the old layout shown in Figure 5-16.

* Poka-yoke devices are mistake-proofing devices that prevent or catch errors or other abnormalities in operating procedures, equipment functions, material or processed parts. For more information about these zero-defect quality control strategies, see Shigeo Shingo, *Zero Quality Control Source Inspection and the Poka-yoke System*, translated by Andrew A. Dillon (Cambridge, Mass., Productivity Press, 1986).

Table 5-1. New Line Organization

Product	Production system	Line	No. of models	Total
Factory No. 1				
Commercial products; spare items; large item SA, CT, GC, ST; etc.	Specialized auto assembly line	Spare line	160 1,350	
Small lot, large variety (SA2, 25), standard shock absorbers	One-piece flow, specialized synchronized line	SA automated assembly line	120	
Large lot, low variety Suspension unit	One-piece flow Mixed line	Blue line	80	
Small lot, large variety Low-pressure gas shock absorber	One-piece flow, specialized synchronized line	HG automated assembly line	20	
Large lot, low variety Low-pressure gas shock absorber	One-piece flow Mixed line	M line	110	
Large lot, low variety Shock absorber with adjustable damping resistance	One-piece flow Mixed line	W line	100	1,940
Factory No. 2				
Small lot, large variety Standard struts (2 types)	One-piece flow, specialized synchronized line	ST-5 line	50	
Small lot, large variety Standard struts (1 type)	"	ST-6 line	10	
Small lot, large variety Standard struts (2 types)	"	ST-8 line	70	
Small lot, large variety Standard struts (1 type)	"	ST-3 line	15	
Small lot, large variety Standard struts (1 type)	"	ST-9 line	35	
Small lot, large variety Standard struts (2 types)	One-piece flow Mixed line	ST-4 line	100	
Small lot, large variety Standard struts (3 types)	"	ST-1 line	60	
Small lot, large variety Specialized struts (3 types)	"	ST-7 line	25	
Small lot, large variety Knuckle-equipped struts	"	ST-2 line	60	
FN-9 air strut	One-piece flow, specialized synchronized line		5	430

Table 5-2. Required Conditions for Processes

Process	Machine or plating process	Welding processes	Assembly processes
Quality	① : Improve rod neck dimensions and stopper for machining rod ends	① : Color code change-over procedure charts and corresponding jigs	① : Color code change-over procedure charts and corresponding jigs
	[2] : Prevent defects in rod screw lengths, etc., improve conveyance jigs	② : Changeover procedure chart	② : Changeover procedure chart
	③ : Take measures to detect rod outer diameters and control chips	[3] : Checkpoints for each product	[3] : Checkpoints for each product
	[4] : Take measures to prevent plating of bearings by photoelectric tubes	④ : Product card (for each parts box)	④ : Product card (for each parts box)
	⑤ : Attach instrument to measure rod's outer diameter	⑤ : Usage range display for gauges	⑤ : Valve assembly components list card for each product
	[6] : Control scattering of chips from processing ends of base shell	[6] : Poka-yoke device to prevent omission of drilling (skipped process)	⑥ : Usage range display for gauges
	[7] : Take measures to prevent inner diameter measurement defects	[7] : Poka-yoke device to prevent reverse attachment of hose brackets	[7] : Length check gauge for tube L
Cost	[1] : Increasing rod polishing speed to 10.5 seconds per rod	① : Reduce bracket weld changeover time from 5 minutes per changeover to 30 seconds per changeover	① : Use machine-readable labels and automate identification operation
	[2] : Speed up rod induction hardening (16.8 - - -> 2 seconds per rod)	[2] : Automate workpiece feed operation for base shell engraver	[2] : Automate by introducing auto-controlled coating chamber
Delivery	① : Improve conveyance by moving the cleaning units for base shells and tubes	[1] : Use stop-and-flow conveyor for mixed one-piece flow	[1] : Connect finishing operations with assembly processes (use coatings for identification and compensation)

○ = autonomous improvement carried out independently on the shop floor

□ = improvement requiring assistance from technical staff

Detailed Design

Following the process-specific flexibility conditions that were listed in step 3, the team implemented various improvements to enhance the flexibility of the processes. Two examples follow.

Mixed Production Improvement: Example 1

This project involved the acceleration of the ST rod induction hardening process and improved flow of goods through the drilling process.

Purpose of improvement. Speeding up induction hardening would boost the production capacity of the existing equipment about 1.5 times (from 65,000 units per month to 97,000 units per month). This cuts costs by bringing in-house production of 32,000 units per month subcontracted to outside vendors at a higher cost.

Description of improvement. (See Figures 5-18 and 5-19.) At the processes shown in Figure 5-18, one operator per shift was able to run the processing of in-house rods in the hardening-tempering-curve straightening-polishing line as well as polishing rods procured from outside vendors. The time for in-house hardened rods was 16.8 seconds per rod due to bottlenecks at the induction hardening and tempering processes. The induction hardening speed was 30 mm per second, and the tempering speed was 15 mm per second.

After improving the coil shape and speeding up the induction hardening and tempering processes, the takt was accelerated 1.5 times (to 11.2 seconds per unit). The hardening speed was also accelerated, from 30 mm per second to 42 mm per second, as was the tempering speed, from 15 mm per second to 24 mm per second.

The polishing process for procured rods was removed and a drilling machine was installed to improve the flow of goods. A

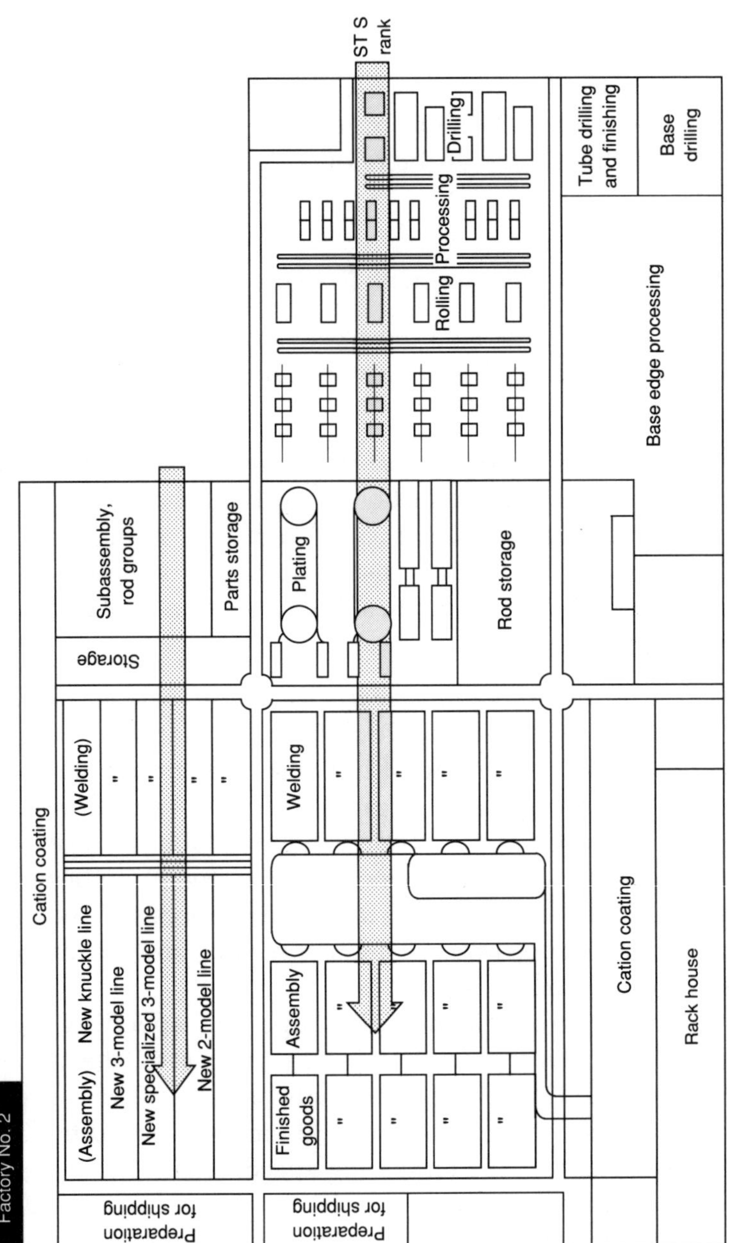

Factory No. 2
Cation coating
(Assembly)
New knuckle line
(Welding)
New 3-model line
New specialized 3-model line
New 2-model line
Preparation for shipping
Storage
Subassembly, rod groups
Parts storage
Plating
Rod storage
Finished goods
Assembly
Welding
Preparation for shipping
Cation coating
Rack house
Rolling
Processing
Drilling
ST S rank
Base edge processing
Tube drilling and finishing
Base drilling

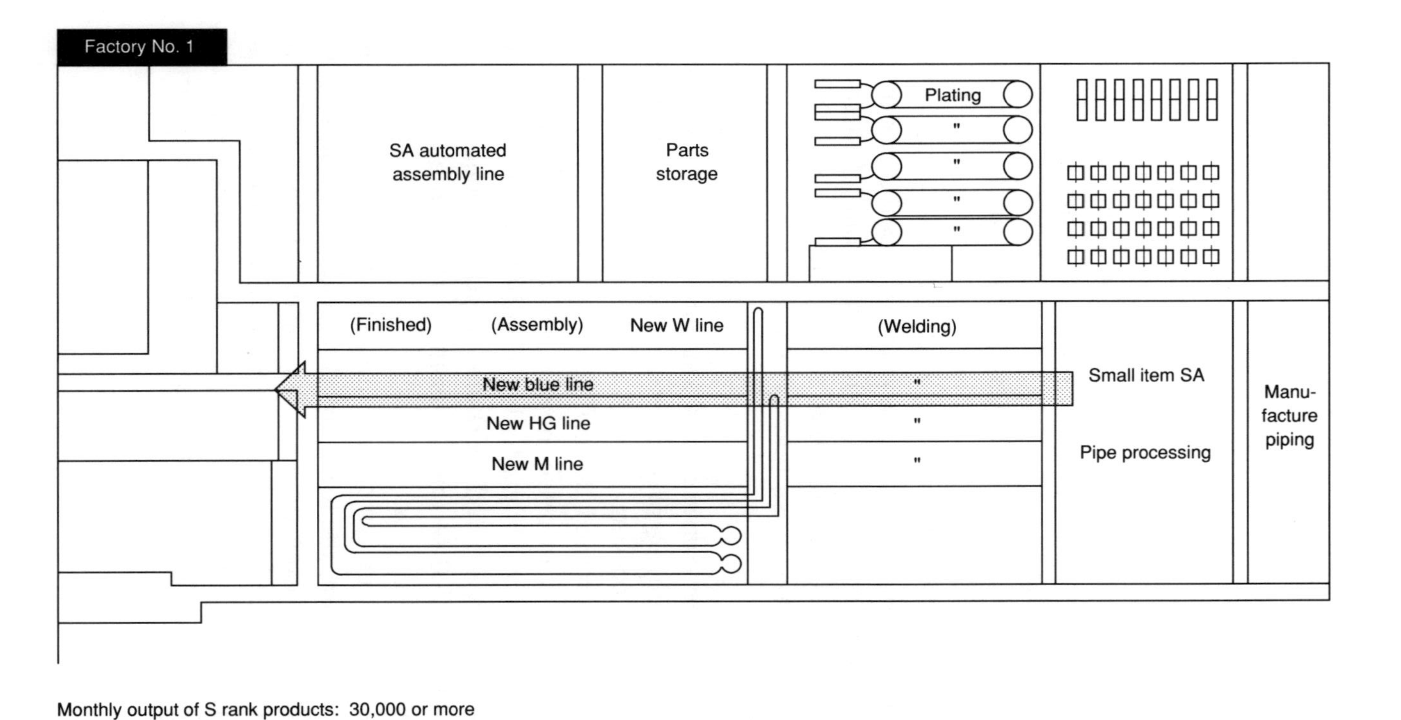

Monthly output of S rank products: 30,000 or more
Monthly output of A rank products: 10,000 to 30,000 or more
Monthly output of B rank products: up to 10,000

Figure 5-17. Basic Layout Plan for Mixed Production Lines

conveyor was installed to link the curve straightening and polishing processes, which lightened the work load imposed by acceleration of other processes.

Effects. After induction hardening and tempering, rods made by the improved production system were compared with rods produced by the previous system. The quality comparison points included surface hardness, variance in effective hardness depth, and bending strength. Both sets of rods met standards.

With respect to cost reduction, ST-reduction savings amounted to $3,300 per month, in-house production savings amounted to $7,600 per month, and energy savings totaled $2,200 per month, for a total savings of $13,100 per month. Rods hardened by outside vendors took a total of three days for drilling, hardening and curve straightening, and polishing. Rods hardened in house take a total of 40 minutes for the same process.

Improvement concerns and key points. Speeding up the induction hardening process means that each workpiece spends less time passing through the heating coil; to compensate, the team increased the coil temperature. This required a 130 percent power increase, however, which was beyond the machine's capacity. They decided finally to increase the coil length so that accelerated workpieces would spend the same amount of time (2.0 seconds) passing through the coil as the slower workpieces.

The shorter the tempering time, the greater the variance in hardness. Again, the team opted to widen the coil to maintain the same heating time (in this case, 1.9 seconds) as that prior to the speed-up improvement. Since there was not enough room to put in a longer coil with the same shape, they installed a coil with vertical pieces on both ends (see Figures 5-18 and 5-19).

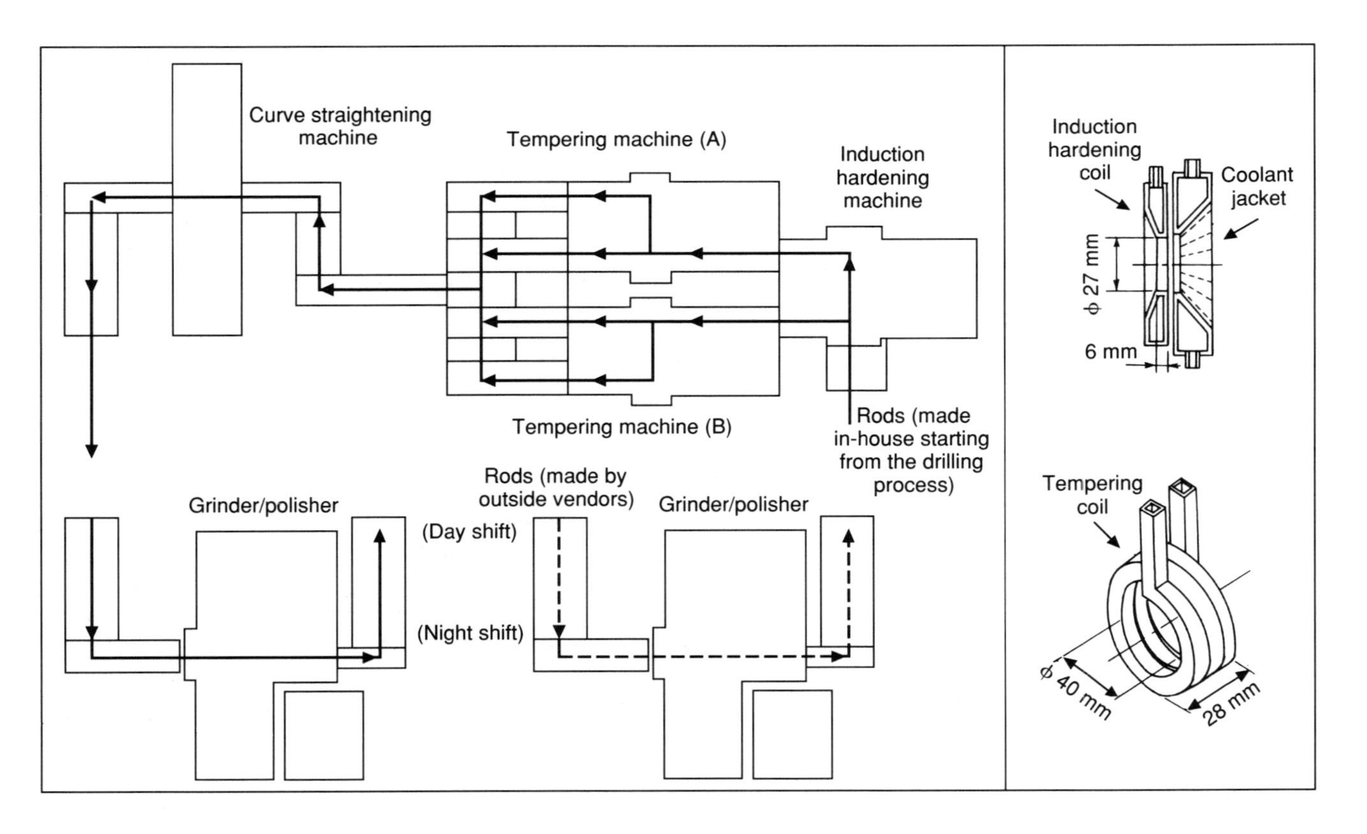

Figure 5-18. Before Improvement

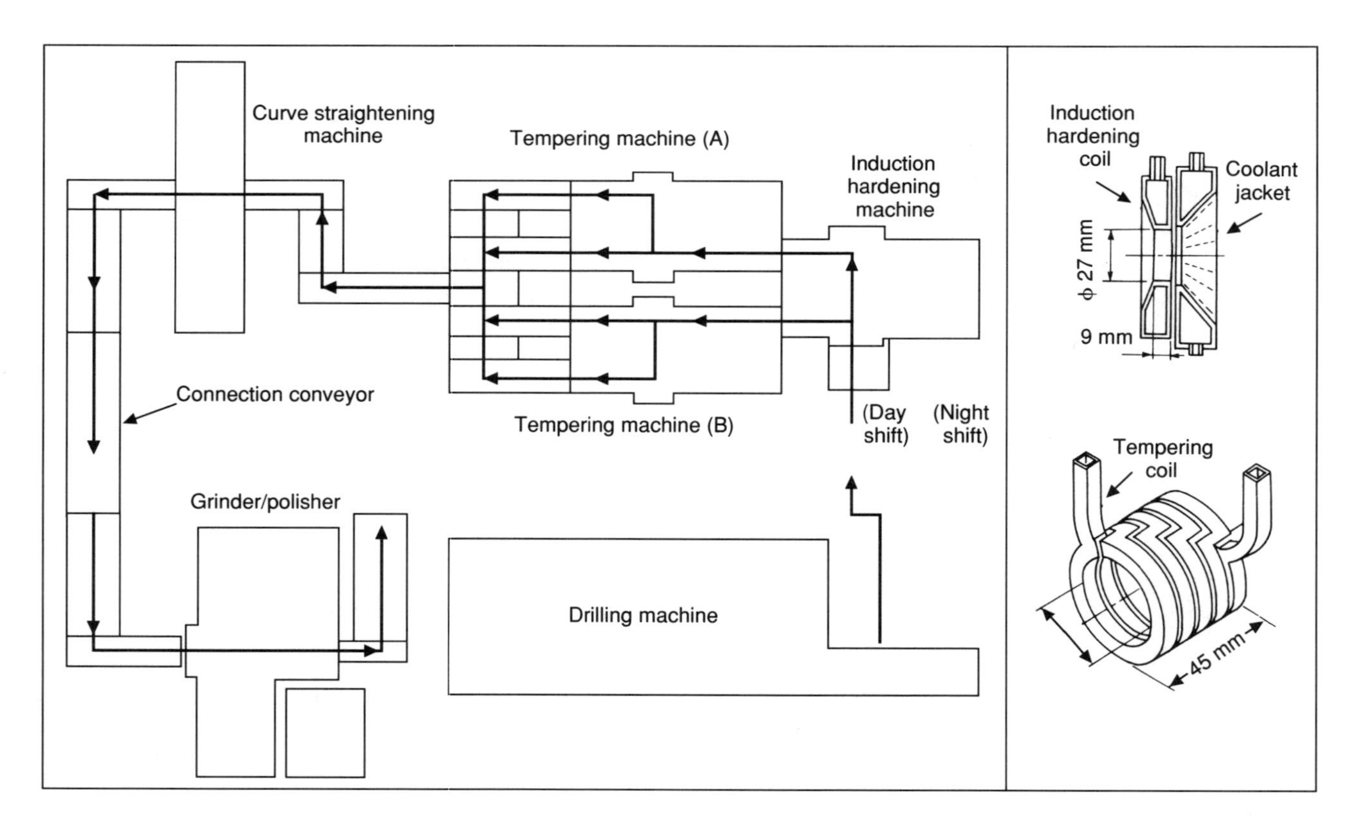

Figure 5-19. After Improvement

Mixed Production Improvement: Example 2

This project involved the establishment of a mixed one-piece flow at welding, coating, and assembly processes using struts and stop-and-flow conveyors.

Purposes of improvement. The improvement was intended to eliminate disruption in the flow of goods, accumulation of work-in-process, conveyance waste, and waste in setting up and removing workpieces. It was also intended to establish a production technology basis supporting small-lot, large-variety production and eventually one-piece flow production.

Description of improvement. See Figures 5-20 and 5-21 below. Before improvement rods were welded, coated, and assembled in large batches and transported by forklift. A large stack of work-in-process was maintained continuously. Installation of the stop-and-flow conveyor made possible a one-piece flow directly to the hanger and provided a direct link to the coating process.

Effects. Before the improvement, products were often dented or damaged in other ways when workpieces were picked up, loaded on carts, or conveyed to the coating chamber. Establishing a one-piece flow eliminated these quality problems.

The new system also resulted in a labor savings of 18 workers and reduced work-in-process inventory before and after the coating chamber from half a day to zero.

Improvement concerns and key points. Different products output by five welding lines are carried in a one-piece flow to a single coating chamber and then are divided up again and fed to five different assembly lines. To achieve this, the paint chamber's conveyors must include mixed-load and location functions. Stop-and-flow conveyors are best for this situation.

After the stop-and-flow conveyor was installed, masking and unmasking workpieces was no longer performed as part of the coating chamber's operations. Instead, those operations were moved to the welding and assembly lines, to be done at the same stations where the workpieces were hooked or unhooked from the conveyor. The welding, coating, and assembly processes are now linked directly, to facilitate product conveyance operations.

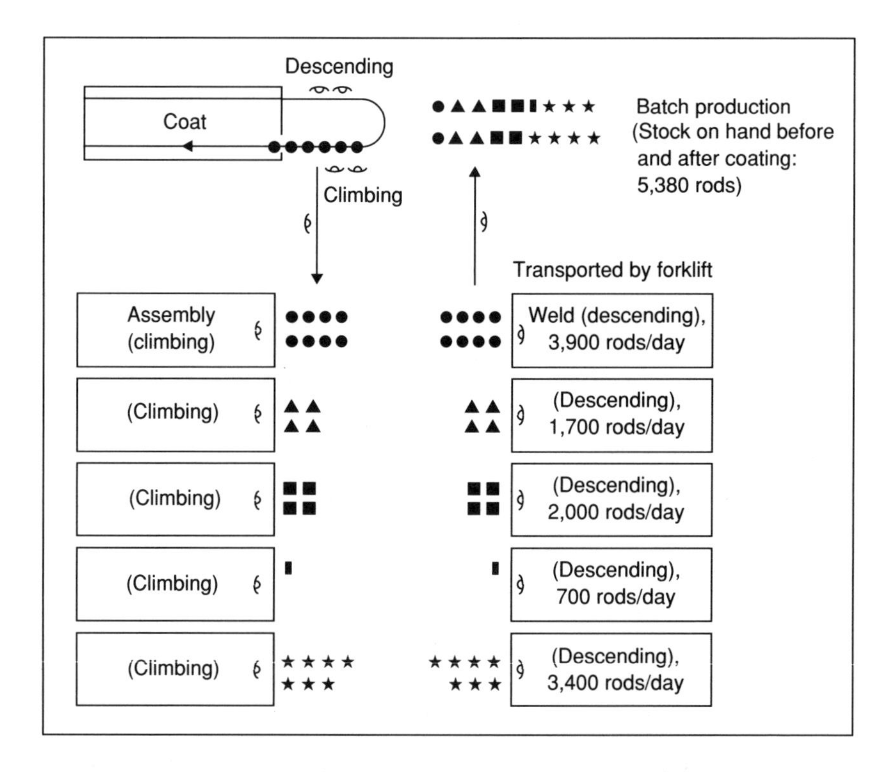

Figure 5-20. Before Improvement

Results

The results of switching to large-variety mixed production using one-piece flow are shown in Figure 5-22; process

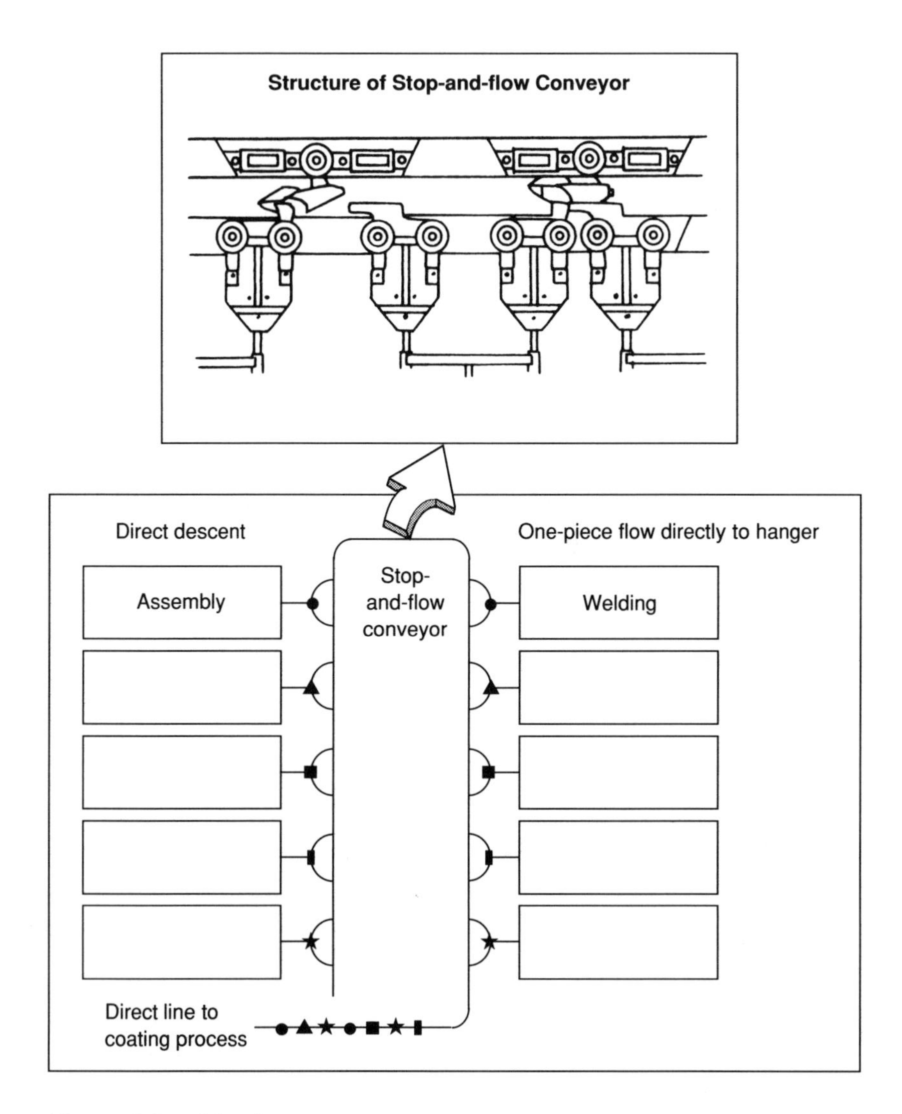

Figure 5-21. After Improvement

defects, work-in-process inventory, and failures dropped significantly, and productivity rose.

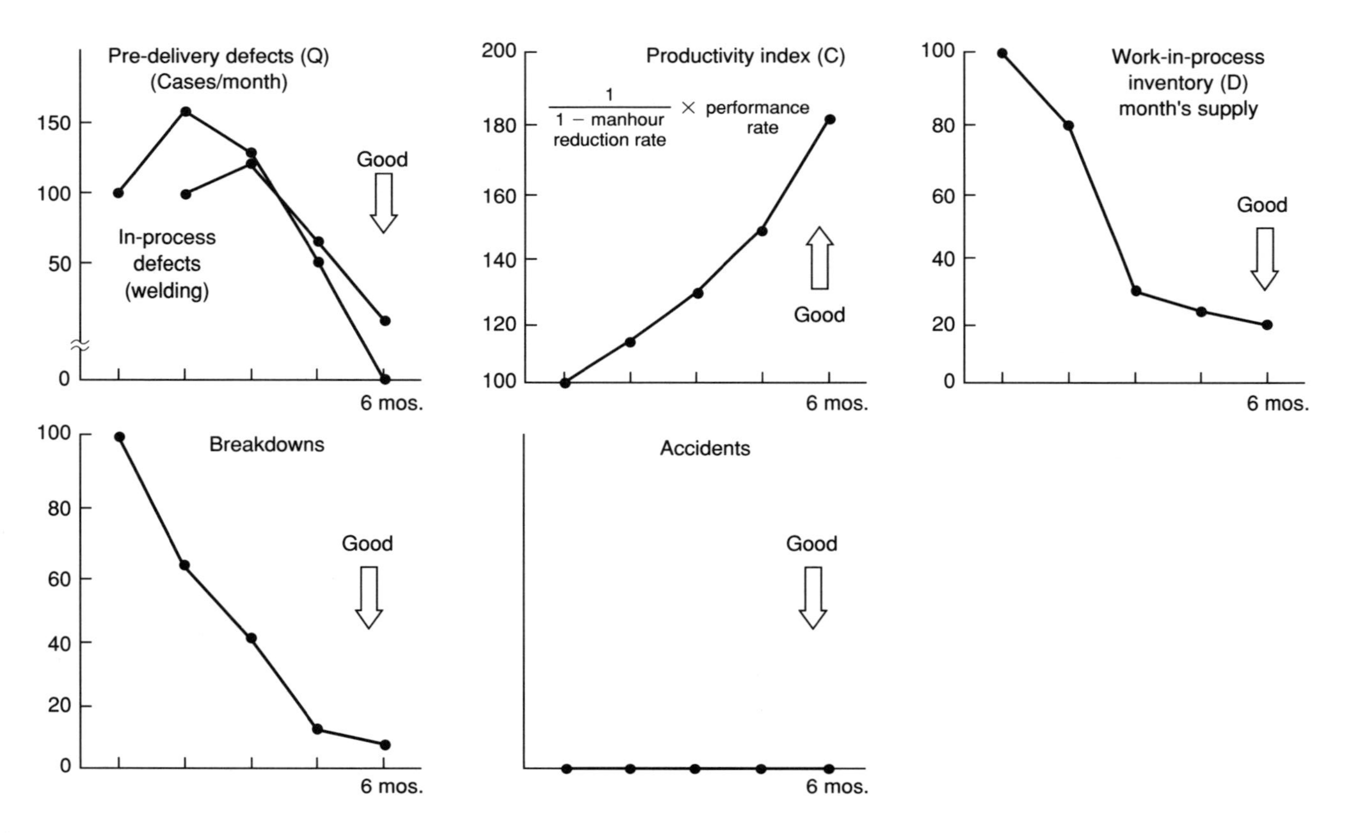

Figure 5-22. Results (Index Displays)

Case Study 5-3: Adding a New Product to an Existing FMS Line

In this case, the same basic approach was taken. Instead of installing a completely new FMS line geared toward a particular group of products, however, this engine factory added new products to an existing FMS line. This presents many more restrictions and makes technical problem-solving all the more important.

Background

The new engine development schedule and existing engine's long-term production schedule are shown below.

a) Startup for new engines

Model \ Business year	1987	1988	1989
A			
B			
C			
D			

Successive development of medium and small engines

b) Existing engines

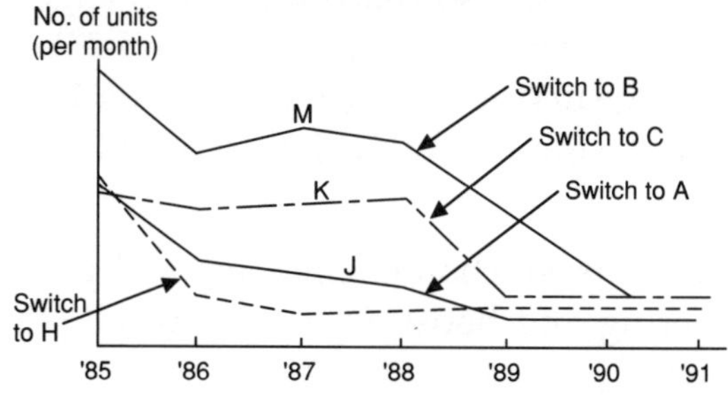

When switching to a new engine model, a big drop occurs in the previous model's production yield. Production of the previous model continues to some extent, however, to use up loaded stock and spare supplies

The basic equipment planning approach for the existing mixed-production line included the following processing and assembly strategies aimed at ensuring efficient production. For example, if processing equipment is centered on NC machines, flexible jig changeover can contribute to a more flexible production line. And if an unmanned transport device is used to supply parts to the assembly line, new parts-supply instructions can be given simply by changing the information cards, which facilitates flexible production.

Plan for Mixing Current Product Assortment with New Products

Table 5-3 shows the plan the design team worked out for mixing a new product into the current assortment of products in the flow. This was then used to determine where bottlenecks and other technical problems might arise and to plan preventive measures.

Product Structure Analysis and Examination of Flow Conditions

A study by the company's engineering division (specifically, the product design department) clarified which mixed-flow problems could be resolved through a change in product specifications and which could not (see Figure 5-23). To promote product designs geared toward easy inclusion in an existing mixed line, the design team decided to adopt the following strategies:

1. Coordinate production engineering and product design engineering and boost exchanges of EP (easy-to-produce) data.
2. Disseminate production equipment specifications.
3. Make processing standards and dimensions of processed items consistent.

Table 5-3. Plan for New and Current Products

Line \ Target models for mixed production	New Engines									Current Engines			
	Under Development				Existing								
	A	B	C	D	E	F	G	H	I	J	K	L	M
1. Block	○	○	—	○	◎	●	—	—	○	×	×	×	×
2. Head	○	○	—	○	◎	●	—	—	●	●	○	●	◎
3. Crank	○	—	—	○	◎	●	—	—	●	●	×	×	◎
4. Cam	○	○	—	○	◎	●	●	—	●	●	×	◎	◎
5. Connecting rods	○	—	—	○	—	—	●	—	●	—	—	—	—
6. Gears	○	○	—	○	◎	●	—	●	—	×	—	—	×
7. Assembly and adjustment	○	○	○	○	◎	○	—	○	×	○	×	×	×

Legend: ●: Existing combination ◎: Now being implemented ○: Combination plan ×: Not combined

4. Standardize positions for workpiece clamps on jigs and pallets.

○: Resolvable ×: Not resolvable

Methods/Parts Location	Methods		Parts			
	Standard-ized	Common settings	Block	Head	Crank	Cam
1. Processing standards	○	—	○	○	○	○
2. Hole depth below screw	○	—	○	○	○	—
3. Standard knock hole position	○	○	○	○	○	—
4. Standard knock hole diameter	○	○	○	○	○	—
5. Fillet R dimension	○	○	—	—	○	—
6. Jig clamp position	—	○	○	○	—	—
7. Journal width dimension	○	—	—	—	○	—
8. Bolt diameter and length under head	○	○	○	○	○	—
9. Bearing diameter	×	—	—	—	×	×
10. Bearing hole diameter	×	—	×	—	—	—

In some cases, the block bearing's hole diameter and the crank's bearing diameter cannot be unified

Figure 5-23. Changing Product Specifications and Mixed Flow Problems

Clarification of Process-specific Conditions

Table 5-4 lists some points concerning bottleneck-prone parts and process conditions that arose during process development. In this example, use of the matrix diagramming method revealed that two block and crank processes required new equipment development and could not be mixed.

Study of Fabrication Methods for Bottleneck Engineering

Figures 5-24 and 5-25 show how this team clarified its bottleneck engineering problems and what specific fabrication methods they chose in response to those problems. For example, in Figure 5-24, fabrication methods were developed for cam- and crank-hole boring at block 6, the process identified as problematic in the previous step.

Detailed Design

At the detailed design stage, team members evaluated the two processes (numbers 6 and 19) that use newly developed equipment units in terms of the five conditions for QA-friendliness. This evaluation helped them predict problem points and study specific countermeasures. They incorporated their findings into the specifications manual. The following design example concerns an enclosed-type boring arbor.

Selection of key parts. (See Figure 5-26.) In this example, team members studied the relationship between required reliability values and key parts of the equipment and identified three areas of concern: arbor friction parts, rotary bearings, and parts linked to the main spindle.

Table 5-4. Trouble-prone Process Conditions

Main Processes	Area of Concern	Tolerance	Equipment Responsiveness	
			Minor Changes	New Development
BLOCK 1. Bore honing	Bore inner diameter	omit	○	—
6. Boring of cam and crank holes	Positioning of cam and crank holes	"	—	●
CRANK, cont'd 18. Cut J strut and width	Width of center J strut	"	—	○
19. Grind journal outer diameter	Journal runout	"	—	●
20. Grind/polish pin	Half-stroke	"	○	—
CONNECTING RODS, cont'd 25. Large and small edge hole boring	Large and small edge hole pitch	"	○	—
26. Large and small edge hole finishing	Large and small edge hole parallelism	"	○	—

Results of analysis: The No. 6 block process (boring of cam and crank holes) and the No. 19 crank process (grind journal outer diameter) require new equipment development and cannot be mixed. Therefore, these are trouble-prone processes.

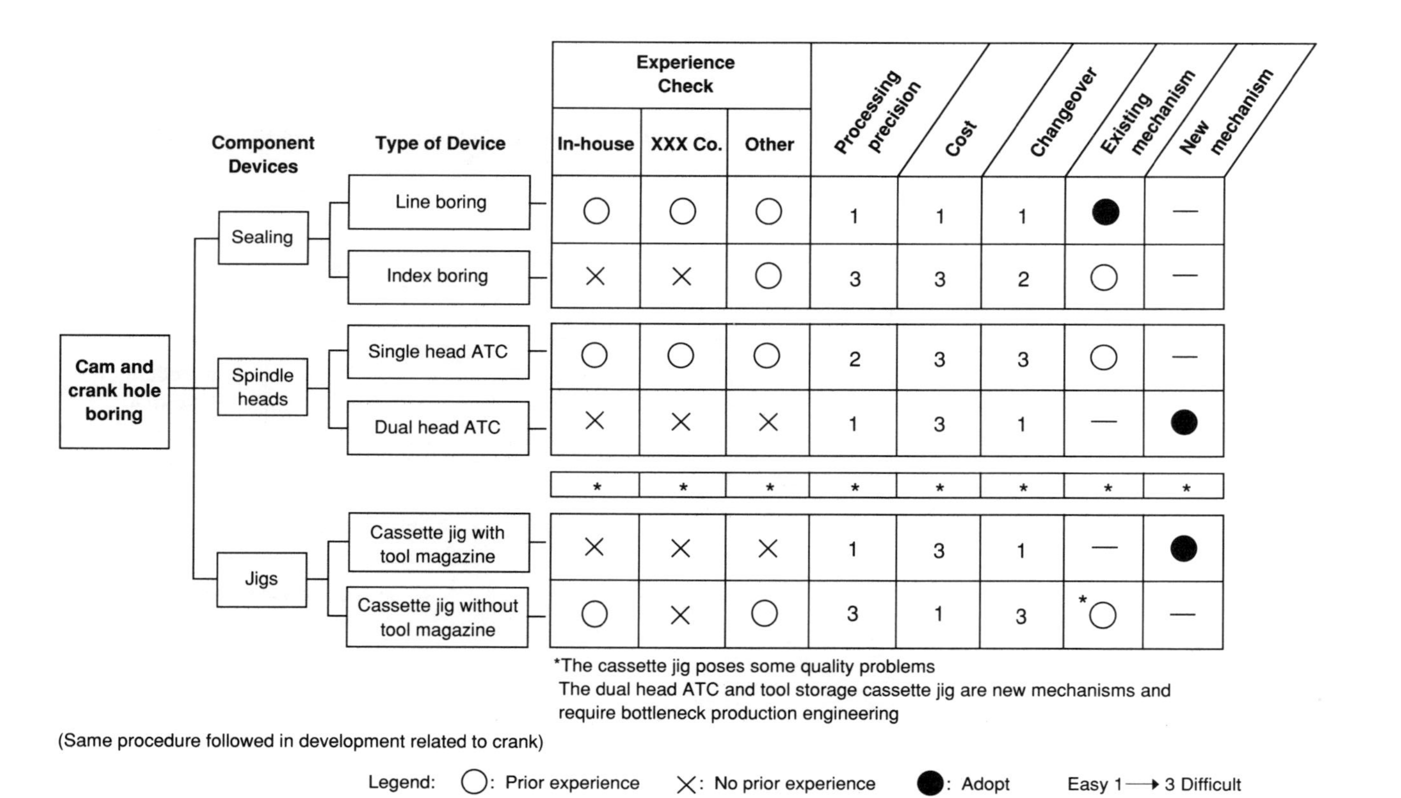

Component Devices	Type of Device	In-house	XXX Co.	Other	Processing precision	Cost	Changeover	Existing mechanism	New mechanism
Sealing	Line boring	○	○	○	1	1	1	●	—
	Index boring	×	×	○	3	3	2	○	—
Spindle heads	Single head ATC	○	○	○	2	3	3	○	—
	Dual head ATC	×	×	×	1	3	1	—	●
		*	*	*	*	*	*	*	*
Jigs	Cassette jig with tool magazine	×	×	×	1	3	1	—	●
	Cassette jig without tool magazine	○	×	○	3	1	3	*○	—

Figure 5-24. Bottleneck Production Technology: Developing New Fabrication Methods at Block No. 6 Process

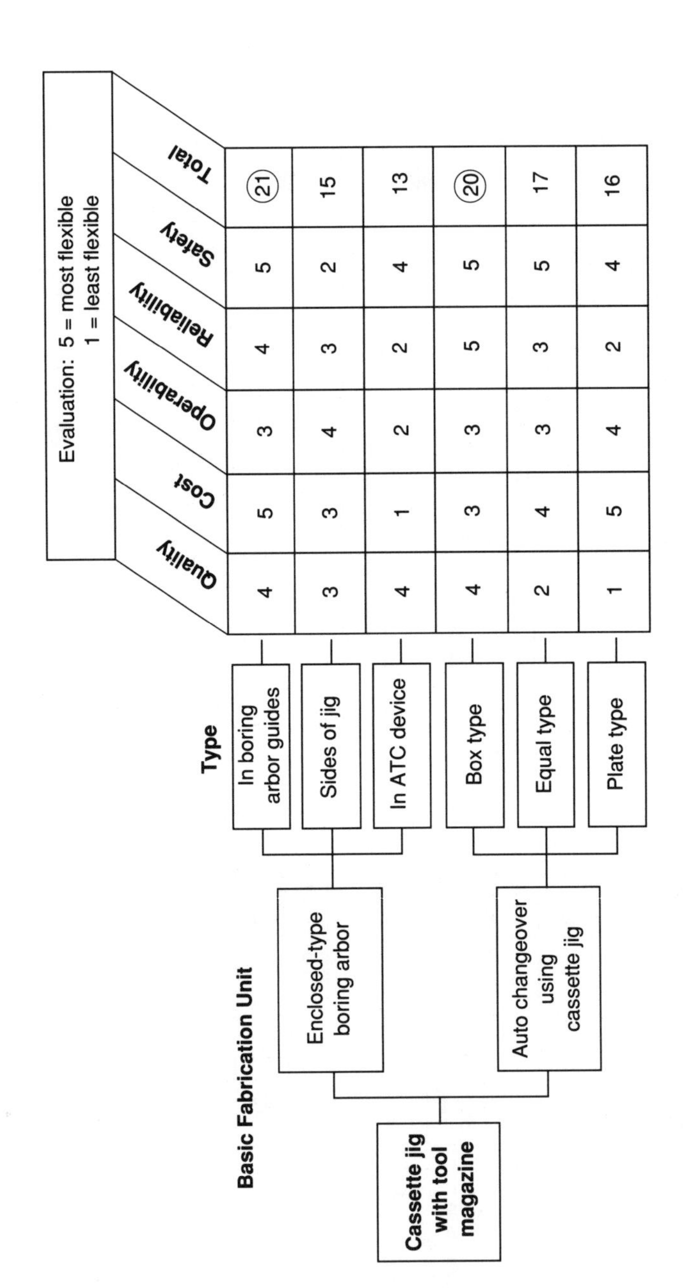

Type	Quality	Cost	Operability	Reliability	Safety	Total
In boring arbor guides	4	5	3	4	5	(21)
Sides of jig	3	3	4	3	2	15
In ATC device	4	1	2	2	4	13
Box type	4	3	3	5	5	(20)
Equal type	2	4	3	3	5	17
Plate type	1	5	4	2	4	16

Figure 5-25. Selecting Methods for New Mechanisms

Selection and evaluation. (See Figure 5-27.) The team then studied the three problem areas in detail, identifying primary and secondary points of concern (highlighted). These were linked with major requirement control items, and the current control methods were evaluated in terms of the five conditions for QA-friendliness. Control items that did not adequately fulfill any of the conditions were singled out for further study.

Selection of countermeasures based on causal analysis. (See Table 5-5.) Here, the team analyzed the problem points identified in step 2 to come up with thorough countermeasures. For example, air pressure (item 5) was hard to check because of the position of the gauge, which had not been specified by the manufacturer. This had an impact on the quality factor of cylindricity. As Table 5-5 shows, the team decided to address the problem at the approval drawing stage of design, to include proper positioning as an item in the manual, and to tell the manufacturer that the gauge needed to be installed at the front of the machine.

DESIGNING FLEXIBILITY FOR SUCCESSIVE MODELS

Today's companies are finding that to stay competitive they must keep pace with fast-changing market needs and respond promptly with appropriate new product development and product model changes. As a result, product life cycles are getting shorter and companies are introducing new products before they can recover the equipment investments made for the current products. The new products also require new equipment investment, which stands the same risk of being unrecoverable. An ongoing trend of equipment investment loss eventually may endanger the company's survival.

Figure 5-28 describes the costs for major welding process equipment that was used for a previous product. As can be seen in the figure, most of the equipment (except for the welding

Relationship between Required Precision and Main Parts

Required Precision	
Item	Tolerance
Outer diameter dimension	omitted
Roundness	"
Cylindricality	"
Surface roughness	"

Evaluation: (Good) 5 ⟷ 1 (Poor)

Key Parts	Impact	Detectability	Occurrence	Importance
1. Arbor friction parts	4	5	3	(60)
2. Arbor drive parts	2	3	3	18
3. Rotary bearings	5	4	4	(60)
4. Parts linked to main spindle	2	2	3	12
5. Part orientation control device	3	3	4	(36)

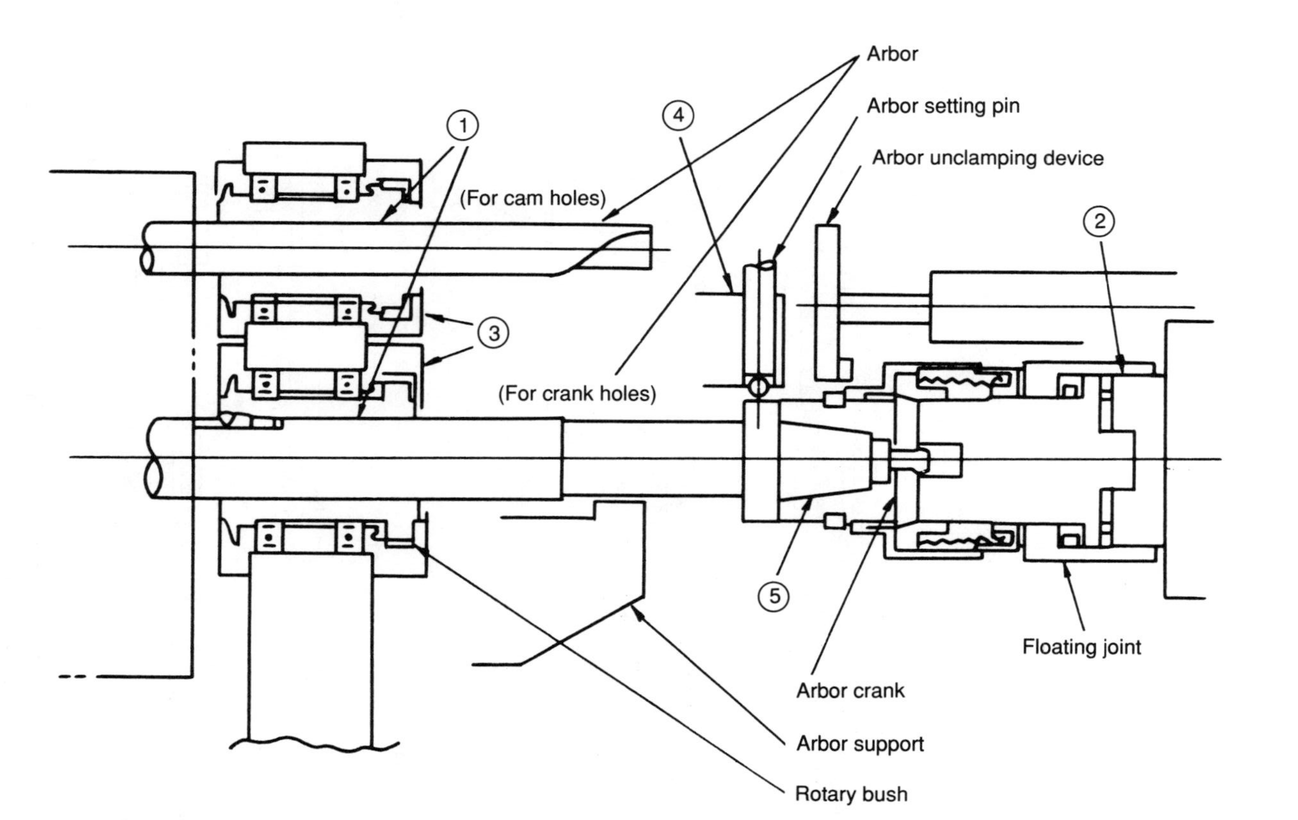

Figure 5-26. Selection of Key Parts

Selection Based on Comparison of Mechanisms and Parts

Points of Concern | Primary Items | Secondary Items

- ① Arbor friction parts
  - Type of material
    - **Extra hard**
    - Steel
  - Precision
    - Balance
      - **Dynamic balance**
      - Static balance
    - **Surface roughness**
    - **Roundness**
    - **Cylindrical shape**
    - **Curved**
  - Environment
    - Cutting debris removal method
      - Air blower
      - **Air purger**
    - Drying method
      - **Dust**
      - Cutting debris
    - Humidifying method
      - Mist
      - Liquid coolant
  - Feed driver
    - Hydraulic
      - Cylinder
      - Load actuator
    - Screws
      - **Boring screws**
      - Trapezoidal screws
  - Feed slide
    - Slide type
      - Metal slide surface
      - Tacite
    - Roller type
      - Ball guide
      - Roller guide
- ③ Rotary bearings
  - Bearing
    - Ball
      - Radial type
      - **Angular type**
    - Roller
      - Tapered
      - Cylindrical
  - Lubrication
    - Liquid
      - Total loss
      - Circulating
      - Supplied
    - Gas
      - Oil mist microfog
    - Paste
      - **Inserted**
- ⑤ Parts linked to main spindle
  - Attach/detach methods
    - Straight
      - Collet
      - Oil bar
    - Tapered
      - **Pull stud**
      - Push-on
    - Cutting debris removal method
      - Air blower
      - **Air purger**

Figure 5-27. Selecting Control Items and Evaluating the Five Conditions for QA-Friendliness

- Development of innovative fabrication methods and equipment that incorporate newly developed technologies or concepts
- Analysis of equipment components to determine whether the equipment structure can be changed to replace specialized functions with general functions

Step 1: Predict future trends in product specifications. Estimate the likelihood of certain changes in product specifications and give those predictions to the product planning and product development departments. When estimating, ask which parts of the product specifications are most prone to change. What kind of functions will be added or eliminated? If product specifications become more complicated — what parts will have the greatest variety?

Step 2: Conduct process analysis to identify bottleneck processes. With the above specification predictions in mind, use process analysis to clarify which processes demand the most flexibility.

Step 3: Predict future investments aimed at bottleneck processes. On the basis of current fabrication methods and equipment, predict about how much equipment investment will be needed to respond to future model changes.

Step 4: Clarify flexibility conditions. Look at individual processes to get a clearer idea of what kind of specification changes will occur and what process-specific flexibility conditions will be required by those changes.

Step 5: Clarify flexibility conditions and process functions. In the context of the process-specific flexibility conditions, clarify what process functions will need to be made flexible. There are two approaches to this, each of which addresses a different situation:

- Situation 1: New fabrication methods and/or equipment must be developed.

- Situation 2: The same fabrication methods can be used, but some parts of the equipment specifications will have to be generalized to make them applicable.

Step 6: Develop fabrication methods and conduct generalization analysis. In situation 1 (step 5) the design results are determined by whether new fabrication methods that provide the various needed functions can be developed. If not, or if situation 2 prevails, you can often solve the problem by changing the equipment specifications to make them more general in application. Figure 5-29 illustrates this latter approach. The procedure follows:

1. Categorize current equipment functions.
2. Enter current equipment specifications on a matrix as "specialized" or "generalized" with regard to each function, then compare the matrix entries (see "A" in Figure 5-29).
3. Explain in detail why parts must be specialized (see "B").
4. Determine the range of generalization based on the relationship between flexibility conditions (from step 5) and functions (see "C").
5. Study alternate plans for generalization.
6. Evaluate the alternate plans to select the most economical (see "D").

Step 7: Determine countermeasures. Use the method described in step 3 to predict and evaluate how investment can be minimized under the selected plan when new products are introduced. If the plan passes this test, adopt it as your final plan.

Case study 5-4 shows how this approach was taken when new fabrication methods had to be developed (situation 1 from step 5).

Steps toward greater use of general purpose equipment:	Equipment / Component / Item	Main body AS plate No. 1 — Side gate: Side frame attachment	Side gate: Opening edge pos.	Side gate: ⋮	Side frame: L-H-W dimensions	Side frame: ⋮	
A Current function	Specialized or general purpose?	SP	G		SP		
B Why specialized?	Due to product shape and/or structure?	○			○		
	Due to limited range of equip. functions?						
C Prerequisites for switching to general purpose	Divide or integrate functions?				○		
	Change dividing points between specialized and general purpose?						
	Change the allocation of functions to simplify equipment?						
	Use alternative functions?						
D Impact on other characteristics	Higher equip. costs?	→	→		↗		
	Lower maintainability?	→	→		↗		
	Higher investment costs for subsequent models?	↗	→		→		

Figure 5-29. Greater Use of General Purpose Equipment

Case Study 5-4: Flexible Design for Successive Models

A factory uses lasers to cut holes in rear floor areas of automobile panels. This hole-cutting process handles workpieces for 50 different vehicle types. The conventional hydraulic punch used before posed the following problems:

- It was difficult to adapt to new specifications or product modifications.
- The equipment was complicated and thus hard to maintain.

As a result, when the factory used this conventional mechanical method, it had to install completely new production equipment for each model change. To resolve this problem, the design team took a fresh look at the fabrication methods used to produce the hole and came up with a completely new approach.

Generally, two types of methods are used for punching holes: cutting methods such as shearing and thermal cutting, and positioning methods such as those for fixed or movable positioning devices. Figure 5-30 shows how different combinations of these methods were evaluated and how their advantages and disadvantages were compared in selecting the best combination. In this schematic example, you can see that of the various shearing methods, the turret punch was more flexible than the specialized hole fabrication methods. Laser cutting was even more attractive.

Figure 5-31 compares the cost of the alternate methods involving robots, turret punches, or carbon dioxide laser. In this case, the laser cutting was the best fabrication method in terms of cost and flexibility toward product designs. While the cost of changing to a new model was similar for both the turret punch and laser cutting methods, the overall cost of the latter was significantly lower.

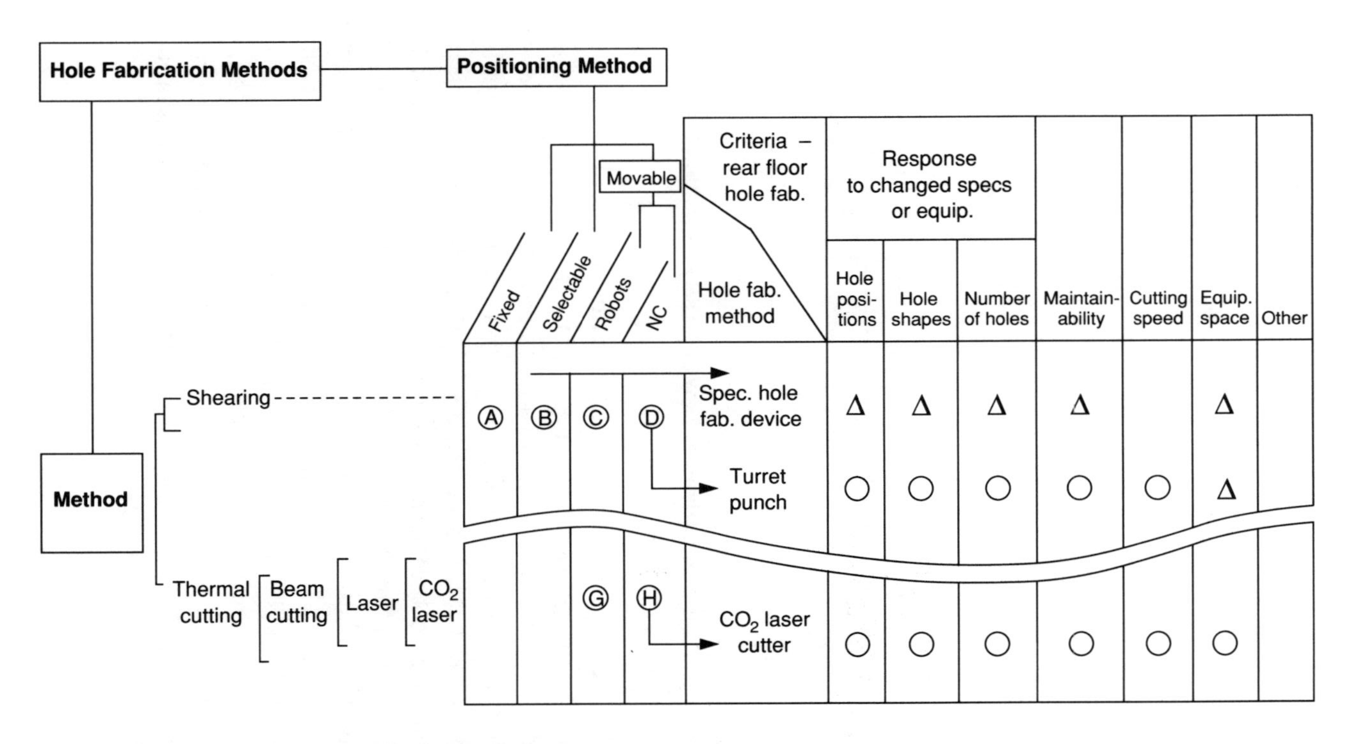

Figure 5-30. Comparative Study of Hole Fabrication Methods

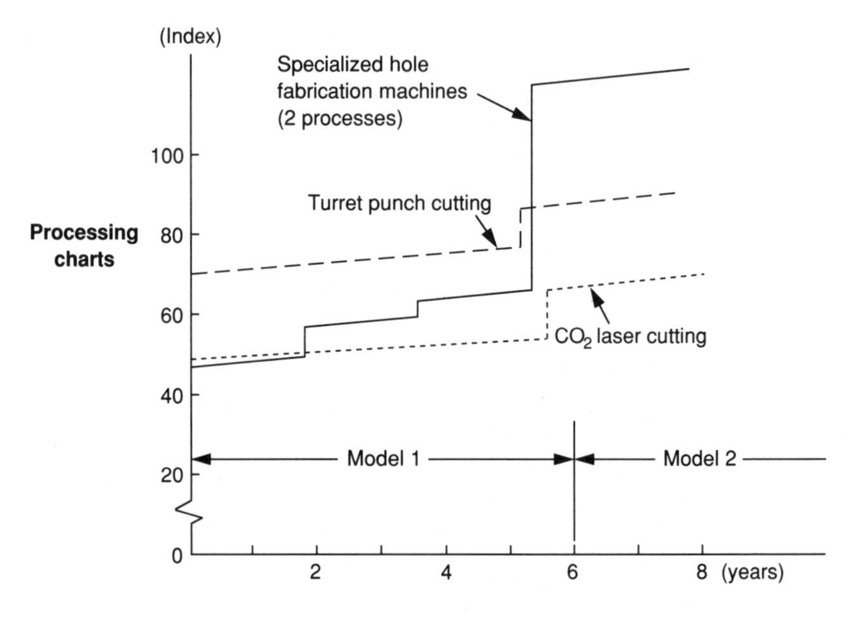

Figure 5-31. Cost Comparison of New and Old Fabrication Methods

This company faced a rather unusual problem, because neither it nor any other Japanese company had any experience introducing this technology to a mass-production line. The team decided to conduct a preliminary study of predicted problems (such as maintaining continuous production in the factory, responding to temperature changes, and removing cutting debris). Using the charting method shown in Figure 5-32, they studied the relationship between abnormalities and equipment components to clarify weak points. Then they drafted measures that would prevent the predicted problems.

The preventive efforts were a success: once installed, the new equipment ran smoothly, without any quality defects or breakdowns. The equipment also proved to be quite adaptable to subsequent product design changes.

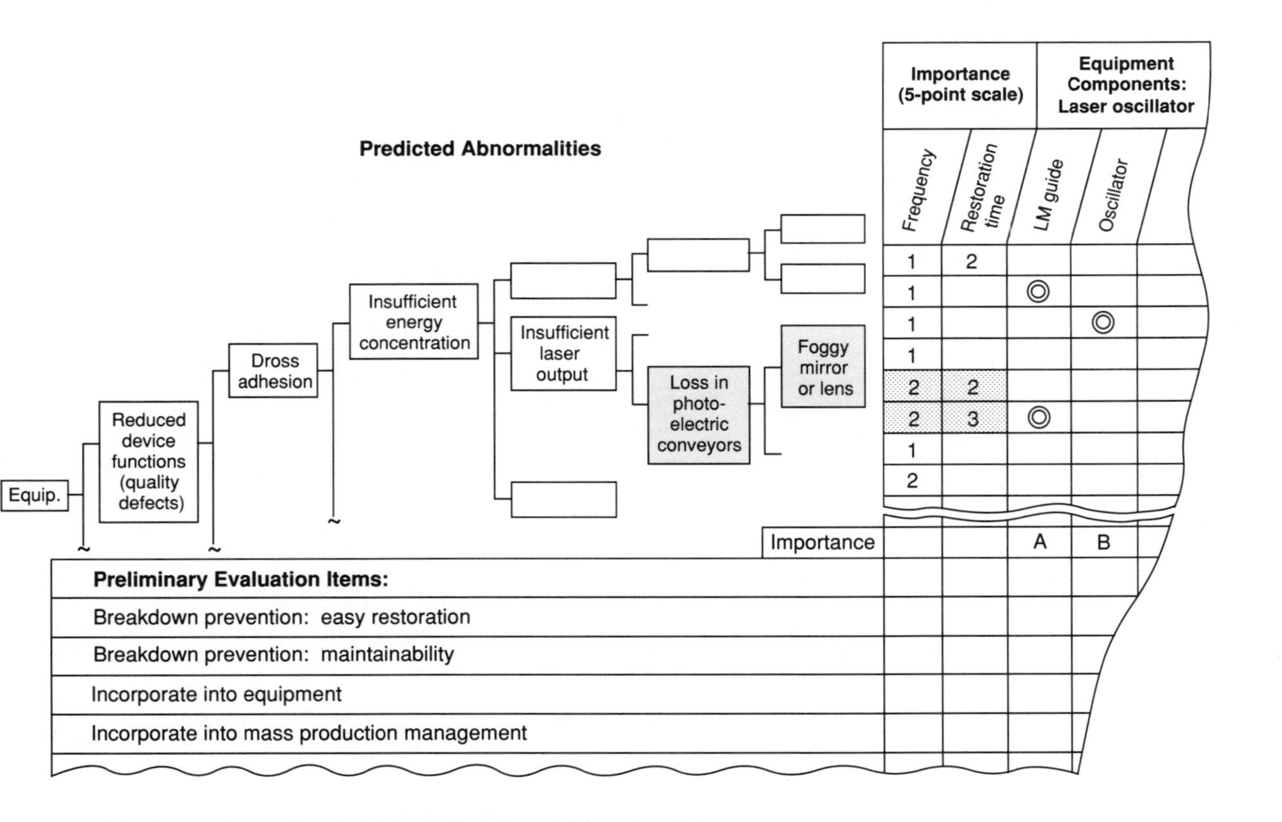

Figure 5-32. Preliminary Evaluation of Predicted Abnormalities

About the Author

Fumio Gotoh is the president of Dymos Consulting Group in Tokyo, which he founded in 1988 after eighteen years as a consultant with the Japan Management Association and the Japan Institute of Plant Maintenance. Mr. Gotoh graduated from Keio University's School of Engineering and joined JMA in 1970, specializing in industrial engineering, quality control, and production control. He began consulting in TPM in 1973, when he developed the now familiar four-phase method for planned maintenance and the seven steps of autonomous maintenance. In 1984 he transferred to JIPM to become head of TPM promotion and was subsequently named head of that organization's TPM research group.

In addition to this book, Mr. Gotoh co-authored *TPM Development Program* and the *TPM Glossary* (in Japanese). His clients have included Tokai Rubber, Topi Kogyo, Toyota Auto Body, Tokiko, Toyota Motors, Shin-Nippon Seitetsu, Bridgestone, Toyoda Automatic Loom Works, Hitachi Seisakusho, Asahi Glass, Tokin Kihin, Fuji Camera/Film, Morinaga Confections, Dainippon Screen, Epson, Fuji Xerox, and Mitsui Mining.

Index

Other Books and AVs on Total Productive Maintenance

Productivity Press publishes and distributes materials on continuous improvement in productivity, quality, customer service, and the creative involvement of all employees. Many of our products are direct source materials from Japan that have been translated into English for the first time and are available exclusively from Productivity. Supplemental products and services include newsletters, conferences, seminars, in-house training and consulting, audio-visual training programs, and industrial study missions. Call 1-800-274-9911 for our free book catalog.

Introduction to TPM
Total Productive Maintenance

by Seiichi Nakajima

Total Productive Maintenance (TPM) combines the American practice of preventive maintenance with the Japanese concepts of total quality control (TQC) and total employee involvement (TEI). The result is an innovative system for equipment maintenance that optimizes effectiveness, eliminates breakdowns, and promotes autonomous operator maintenance through day-to-day activities. This book summarizes the steps involved in TPM and provides case examples from several top Japanese plants.
ISBN 0-915299-23-2 / 149 pages / $39.95 / Order code ITPM-BK

TPM Development Program
Implementing Total Productive Maintenance

edited by Seiichi Nakajima

This book outlines a three-year program for systematic TPM development and implementation. It describes in detail the five principal developmental activities of TPM:

1. Systematic elimination of the six big equipment related losses through small group activities
2. Autonomous maintenance (by operators)
3. Scheduled maintenance for the maintenance department
4. Training in operation and maintenance skills
5. Comprehensive equipment management from the design stage

ISBN 0-915299-37-2 / 428 pages / $85.00 / Order code DTPM-BK

Training for TPM
A Manufacturing Success Story

edited by Nachi-Fujikoshi

A detailed case study of TPM implementation at a world-class manufacturer of bearings, precision machine tools, dies, industrial equipment, and robots. In 1984, two and a half years after beginning implementation, the company was awarded Japan's prestigious PM Prize for its program. Here's a detailed account of their improvement activities — and an impressive model for yours.
ISBN 0-915299-34-8 / 320 pages / $59.95 / Order code CTPM-BK

Total Manufacturing Management
Production Organization for the 1990s

by Giorgio Merli

One of Italy's leading consultants discusses the implementation of Just-In-Time and related methods (including QFD and TPM) in Western corporations. The author does not approach JIT from a mechanistic orientation aimed simply at production efficiency. Rather, he discusses JIT from the perspective of industrial strategy and as an overall organizational model. Here's a sophisticated program for organizational reform that shows how JIT can be applied even in types of production that have often been neglected in the West, including custom work.
ISBN 0-915299-58-5 / 224 pages / $39.95 / Order code TMM-BK

The Basic Concept of Total Productive Maintenance

by Seiichi Nakajima

The more you learn about the seemingly limitless benefits of Total Productive Maintenance (TPM), the more you will want a thorough and accurate source of information on its methodology and applications. This three-part video set brings years of TPM experience in Japanese industry directly to your whole company in an interesting and information-packed presentation. It describes a 12-step program for introducing and developing TPM in your company and includes case studies with comments by the leaders of companies that have won Japan's TPM prize (awarded those who have demonstrated excellence in TPM implementation).
3 videotapes (80 min.) / $1,100.00 / Order YASTPM-BK

COMPLETE LIST OF TITLES FROM PRODUCTIVITY PRESS

Akao, Yoji (ed.). **Quality Function Deployment: Integrating Customer Requirements into Product Design**
ISBN 0-915299-41-0 / 1990 / 387 pages / $ 75.00 / order code QFD

Akiyama, Kaneo. **Function Analysis: Systematic Improvement of Quality and Performance**
ISBN 0-915299-81-X / 1991 / 288 pages / $59.95 / order code FA

Asaka, Tetsuichi and Kazuo Ozeki (eds.). **Handbook of Quality Tools: The Japanese Approach**
ISBN 0-915299-45-3 / 1990 / 336 pages / $59.95 / order code HQT

Belohlav, James A. **Championship Management: An Action Model for High Performance**
ISBN 0-915299-76-3 / 1990 / 265 pages / $29.95 / order code CHAMPS

Birkholz, Charles and Jim Villella. **The Battle to Stay Competitive: Changing the Traditional Workplace**
ISBN 0-915299-96-8 / 1991 / 110 pages paper / $9.95 /order code BATTLE

Christopher, William F. **Productivity Measurement Handbook**
ISBN 0-915299-05-4 / 1985 / 680 pages / $137.95 / order code PMH

D'Egidio, Franco. **The Service Era: Leadership in a Global Environment**
ISBN 0-915299-68-2 / 1990 / 165 pages / $29.95 / order code SERA

Ford, Henry. **Today and Tomorrow**
ISBN 0-915299-36-4 / 1988 / 286 pages / $24.95 / order code FORD

Fukuda, Ryuji. **CEDAC: A Tool for Continuous Systematic Improvement**
ISBN 0-915299-26-7 / 1990 / 144 pages / $49.95 / order code CEDAC

Fukuda, Ryuji. **Managerial Engineering: Techniques for Improving Quality and Productivity in the Workplace** (rev.)
ISBN 0-915299-09-7 / 1986 / 208 pages / $39.95 / order code ME

Gotoh, Fumio. **Equipment Planning for TPM: Maintenance Prevention Design**
ISBN 0-915299-77-1 / 1991 / 272 pages / $75.00 / order code ETPM

Grief, Michel. **The Visual Factory: Building Participation Through Shared Information**
ISBN 0-915299-67-4 / 1991 / 320 pages / $49.95 / order code VFAC

Hatakeyama, Yoshio. **Manager Revolution! A Guide to Survival in Today's Changing Workplace**
ISBN 0-915299-10-0 / 1986 / 208 pages / $24.95 / order code MREV

Hirano, Hiroyuki. **JIT Factory Revolution: A Pictorial Guide to Factory Design of the Future**
ISBN 0-915299-44-5 / 1989 / 227 pages / $49.95 / order code JITFAC

Hirano, Hiroyuki. **JIT Implementation Manual: The Complete Guide to Just-In-Time Manufacturing**
ISBN 0-915299-66-6 / 1990 / 1006 pages / $2500.00 / order code HIRANO

Horovitz, Jacques. **Winning Ways: Achieving Zero-Defect Service**
ISBN 0-915299-78-X / 1990 / 165 pages / $24.95 / order code WWAYS

Ishiwata, Junichi. **IE for the Shop Floor 1: Productivity through Process Analysis**
ISBN 0-915299-82-8 / 1991 / 208 pages / $39.95 / order code SHOPF1

Japan Human Relations Association (ed.). **The Idea Book: Improvement Through TEI (Total Employee Involvement)**
ISBN 0-915299-22-4 / 1988 / 232 pages / $49.95 / order code IDEA

Japan Human Relations Association (ed.). **The Service Industry Idea Book: Employee Involvement in Retail and Office Improvement**
ISBN 0-915299-65-8 / 1990 / 294 pages / $49.95 / order code SIDEA

Japan Management Association (ed.). **Kanban and Just-In-Time at Toyota: Management Begins at the Workplace** (rev.), Translated by David J. Lu
ISBN 0-915299-48-8 / 1989 / 224 pages / $36.50 / order code KAN

Japan Management Association and Constance E. Dyer. **The Canon Production System: Creative Involvement of the Total Workforce**
ISBN 0-915299-06-2 / 1987 / 251 pages / $36.95 / order code CAN

Jones, Karen (ed.). **The Best of TEI: Current Perspectives on Total Employee Involvement**
ISBN 0-915299-63-1 / 1989 / 502 pages / $175.00 / order code TEI

JUSE. **TQC Solutions: The 14-Step Process**
ISBN 0-915299-79-8 / 1991 / 416 pages / 2 volumes / $120.00 / order code TQCS

Kanatsu, Takashi. **TQC for Accounting: A New Role in Companywide Improvement**
ISBN 0-915299-73-9 / 1991 / 244 pages / $45.00 / order code TQCA

Karatsu, Hajime. **Tough Words For American Industry**
ISBN 0-915299-25-9 / 1988 / 178 pages / $24.95 / order code TOUGH

Karatsu, Hajime. **TQC Wisdom of Japan: Managing for Total Quality Control**, Translated by David J. Lu
ISBN 0-915299-18-6 / 1988 / 136 pages / $34.95 / order code WISD

Kato, Kenichiro. **I.E. for the Shop Floor 2: Productivity through Motion Study**
ISBN 1-56327-000-5 / 1991 / 224 pages / $39.95 / order code SHOPF2

Kaydos, Will. **Measuring, Managing, and Maximizing Performance**
ISBN 0-915299- 98-4 / 1991 / 208 pages / $34.95 / order code MMMP

Kobayashi, Iwao. **20 Keys to Workplace Improvement**
ISBN 0-915299-61-5 / 1990 / 264 pages / $34.95 / order code 20KEYS

Lu, David J. **Inside Corporate Japan: The Art of Fumble-Free Management**
ISBN 0-915299-16-X / 1987 / 278 pages / $24.95 / order code ICJ

Maskell, Brian H. **Performance Measurement for World Class Manufacturing: A Model for American Companies**
ISBN 0-915299-99-2 / 1991 / 448 pages / 45.00 / order code PERFM

Merli, Giorgio. **Total Manufacturing Management: Production Organization for the 1990s**
ISBN 0-915299-58-5 / 1990 / 224 pages / $39.95 / order code TMM

Mizuno, Shigeru (ed.). **Management for Quality Improvement: The 7 New QC Tools**
ISBN 0-915299-29-1 / 1988 / 324 pages / $59.95 / order code 7QC

Monden, Yasuhiro and Michiharu Sakurai (eds.). **Japanese Management Accounting: A World Class Approach to Profit Management**
ISBN 0-915299-50-X / 1990 / 568 pages / $59.95 / order code JMACT

Nachi-Fujikoshi (ed.). **Training for TPM: A Manufacturing Success Story**
ISBN 0-915299-34-8 / 1990 / 272 pages / $59.95 / order code CTPM

Nakajima, Seiichi. **Introduction to TPM: Total Productive Maintenance**
ISBN 0-915299-23-2 / 1988 / 149 pages / $39.95 / order code ITPM

Nakajima, Seiichi. **TPM Development Program: Implementing Total Productive Maintenance**
ISBN 0-915299-37-2 / 1989 / 428 pages / $85.00 / order code DTPM

Nikkan Kogyo Shimbun, Ltd./Factory Magazine (ed.). **Poka-yoke: Improving Product Quality by Preventing Defects**
ISBN 0-915299-31-3 / 1989 / 288 pages / $59.95 / order code IPOKA

NKS/Esme McTighe (ed.). **Factory Management Notebook Series: Mixed Model Production**
ISBN 0-915299-97-6 / 1991 / 184 / $175.00 / order code N1-MM

NKS/Esme McTighe (ed.). **Factory Management Notebook Series: Visual Control Systems: Visual Control Systems**
ISBN 0-915299-54-2 / 1991 / 194 pages / $175.00 / order code N1-VCS

NKS/Esme McTighe (ed.). **Factory Management Notebook Series: Mixed Model Production**
ISBN 0-0-56327-002-1 / 1991 / 200 pages / $175.00 / order code N1-AA

Ohno, Taiichi. **Toyota Production System: Beyond Large-scale Production**
ISBN 0-915299-14-3 / 1988 / 162 pages / $39.95 / order code OTPS

Ohno, Taiichi. **Workplace Management**
ISBN 0-915299-19-4 / 1988 / 165 pages / $34.95 / order code WPM

Ohno, Taiichi and Setsuo Mito. **Just-In-Time for Today and Tomorrow**
ISBN 0-915299-20-8 / 1988 / 208 pages / $34.95 / order code OMJIT

Perigord, Michel. **Achieving Total Quality Management: A Program for Action**
ISBN 0-915299-60-7 / 1991 / 384 pages / $45.00 / order code ACHTQM

Psarouthakis, John. **Better Makes Us Best**
ISBN 0-915299-56-9 / 1989 / 112 pages / $16.95 / order code BMUB

Robinson, Alan. **Continuous Improvement in Operations: A Systematic Approach to Waste Reduction**
ISBN 0-915299-51-8 / 1991 / 416 pages / $34.95 / order code ROB2-C

Robson, Ross (ed.). T**he Quality and Productivity Equation: American Corporate Strategies for the 1990s**
ISBN 0-915299-71-2 / 1990 / 558 pages / $29.95 / order code QPE

Shetty, Y.K and Vernon M. Buehler (eds.). **Competing Through Productivity and Quality**
ISBN 0-915299-43-7 / 1989 / 576 pages / $39.95 / order code COMP

Shingo, Shigeo. **Non-Stock Production: The Shingo System for Continuous Improvement**
ISBN 0-915299-30-5 / 1988 / 480 pages / $75.00 / order code NON

Shingo, Shigeo. **A Revolution In Manufacturing: The SMED System**, Translated by Andrew P. Dillon
ISBN 0-915299-03-8 / 1985 / 383 pages / $70.00 / order code SMED

Shingo, Shigeo. **The Sayings of Shigeo Shingo: Key Strategies for Plant Improvement**, Translated by Andrew P. Dillon
ISBN 0-915299-15-1 / 1987 / 208 pages / $39.95 / order code SAY

Shingo, Shigeo. **A Study of the Toyota Production System from an Industrial Engineering Viewpoint** (rev.)
ISBN 0-915299-17-8 / 1989 / 293 pages / $39.95 / order code STREV

Shingo, Shigeo. **Zero Quality Control: Source Inspection and the Poka-yoke System**,Translated by Andrew P. Dillon
ISBN 0-915299-07-0 / 1986 / 328 pages / $70.00 / order code ZQC

Shinohara, Isao (ed.). **New Production System: JIT Crossing Industry Boundaries**
ISBN 0-915299-21-6 / 1988 / 224 pages / $34.95 / order code NPS

Sugiyama, Tomo. **The Improvement Book: Creating the Problem-Free Workplace**
ISBN 0-915299-47-X / 1989 / 236 pages / $49.95 / order code IB

Suzue, Toshio and Akira Kohdate. **Variety Reduction Program (VRP): A Production Strategy for Product Diversification**
ISBN 0-915299-32-1 / 1990 / 164 pages / $59.95 / order code VRP

Tateisi, Kazuma. **The Eternal Venture Spirit: An Executive's Practical Philosophy**
ISBN 0-915299-55-0 / 1989 / 208 pages/ $19.95 / order code EVS

Yasuda, Yuzo. **40 Years, 20 Million Ideas: The Toyota Suggestion System**
ISBN 0-915299-74-7 / 1991 / 210 pages / $39.95 / order code 4020

Audio-Visual Programs

Japan Management Association. **Total Productive Maintenance: Maximizing Productivity and Quality**
ISBN 0-915299-46-1 / 167 slides / 1989 / $749.00 / order code STPM
ISBN 0-915299-49-6 / 2 videos / 1989 / $749.00 / order code VTPM

Shingo, Shigeo. **The SMED System**, Translated by Andrew P. Dillon
ISBN 0-915299-11-9 / 181 slides / 1986 / $749.00 / order code S5
ISBN 0-915299-27-5 / 2 videos / 1987 / $749.00 / order code V5

Shingo, Shigeo. **The Poka-yoke System**, Translated by Andrew P. Dillon
ISBN 0-915299-13-5 / 235 slides / 1987 / $749.00 / order code S6
ISBN 0-915299-28-3 / 2 videos / 1987 / $749.00 / order code V6

Returns of AV programs willl be accepted for incorrect or damaged shipments only.

TO ORDER: Write, phone, or fax Productivity Press, Dept. BK, P.O. Box 3007, Cambridge, MA 02140, phone 1-800-274-9911, fax 617-864-6286. Send check or charge to your credit card (American Express, Visa, MasterCard accepted).

U.S. ORDERS: Add $5 shipping for first book, $2 each additional for UPS surface delivery. CT residents add 8% and MA residents 5% sales tax. For each AV program that you order, add $5 for programs with 1 or 2 tapes, and $12 for programs with 3 or more tapes.

INTERNATIONAL ORDERS: Write, phone, or fax for quote and indicate shipping method desired. Pre-payment in U.S. dollars must accompany your order (checks must be drawn on U.S. banks). When quote is returned with payment, your order will be shipped promptly by the method requested.

NOTE: Prices subject to change without notice.